高等院校经济管理类专业"互联网+"创新规划教材

# DATA MINING AND MANAGEMENT DECISION MAKING

# 数据挖掘与管理决策
（双语版）

陈欣 ◎主编

北京大学出版社
PEKING UNIVERSITY PRESS

## 内 容 简 介

本书以数据挖掘经典算法及其在管理决策中的应用为研究对象进行编写，内容包括数据挖掘的发展背景、分类算法、聚类算法和关联规则算法、人工神经网络算法及数据挖掘在管理决策中的应用等，基本涵盖了数据挖掘技术中的经典算法。通过介绍在不同管理决策领域中数据挖掘技术的典型案例，将理论学习和实际应用有机地结合起来，有助于不同专业背景的学生更好地学习和掌握数据挖掘的基本技术和发展动态。

本书可作为高等院校经管类和理工类专业高年级本科生或研究生的入门教材，也可作为高等院校数据挖掘方向双语课程建设的参考教材，还可作为对数据挖掘技术感兴趣的读者的参考书籍。

**图书在版编目(CIP)数据**

数据挖掘与管理决策：双语版/陈欣主编．—北京：北京大学出版社，2022.8
高等院校经济管理类专业"互联网+"创新规划教材
ISBN 978-7-301-33281-8

Ⅰ.①数… Ⅱ.①陈… Ⅲ.①数据采掘—双语教学—高等学校—教材 Ⅳ.①TP311.131

中国版本图书馆 CIP 数据核字(2022)第 151656 号

| | |
|---|---|
| 书　　　名 | 数据挖掘与管理决策（双语版）<br>SHUJU WAJUE YU GUANLI JUECE (SHUANGYU BAN) |
| 著作责任者 | 陈　欣　主编 |
| 责 任 编 辑 | 王显超　李娉婷 |
| 数 字 编 辑 | 金常伟 |
| 标 准 书 号 | ISBN 978-7-301-33281-8 |
| 出 版 发 行 | 北京大学出版社 |
| 地　　　址 | 北京市海淀区成府路 205 号　100871 |
| 网　　　址 | http://www.pup.cn　新浪微博：@北京大学出版社 |
| 电 子 信 箱 | pup_6@163.com |
| 电　　　话 | 邮购部 010-62752015　发行部 010-62750672　编辑部 010-62750667 |
| 印 刷 者 | 北京鑫海金澳胶印有限公司 |
| 经 销 者 | 新华书店 |
| | 787 毫米×1092 毫米　16 开本　12 印张　341 千字<br>2022 年 8 月第 1 版　2022 年 8 月第 1 次印刷 |
| 定　　　价 | 35.00 元 |

未经许可，不得以任何方式复制或抄袭本书之部分或全部内容。
**版权所有，侵权必究**
举报电话：010-62752024　电子信箱：fd@pup.pku.edu.cn
图书如有印装质量问题，请与出版部联系，电话：010-62756370

# 前　言

　　进入 21 世纪以来，随着海量数据库系统的建立和互联网的普及使用，"大数据"时代已然到来，与之密切相关的数据挖掘技术也得以迅猛发展，并在工业界和学术界引发了广泛关注。数据挖掘技术可用于分析和发现隐藏在海量数据背后的关系、规则和模式等，为商业竞争、企业生产和管理、政府部门决策以及科学探索等提供有用信息支持，已成为计算机科学及相关领域的热门技术。

　　数据挖掘是由数据库、统计学、机器学习、模式识别以及决策支持系统等多个范畴的理论和技术交叉融合而形成的一门综合性学科。它所涉及的知识点范围非常广，很难用一本书囊括所有相关技术。对一本教科书来说，更是难以做到这一点。事实上，作为数据挖掘的初学者，也没有必要全部掌握这些技术，可以先从一些常用的、基础的数据挖掘方法入手，再逐步扩展到更高级的算法学习。这是本书写作的一个重要出发点。

　　本书共分为 6 章。第 1 章是本书的导论，系统地介绍了数据挖掘基本思想所需的信息，如定义、实现过程和技术、在决策中的应用以及数据挖掘的发展等，本章主要是为读者提供数据挖掘概念的基本背景知识；第 2 章～第 5 章，介绍数据挖掘中的经典和常用技术，具体包括用于分类、聚类和关联规则的主要算法以及人工神经网络算法，它们是其他众多数据挖掘课程中最常见的算法。这些章节内容还包括相应算法的用法说明、例子以及用 Python 或 R 语言实现的代码，有助于读者理解和掌握数据挖掘的基本概念、原理和方法。第 6 章，介绍数据挖掘在管理决策中的应用，使读者在理论学习的基础上，通过行业应用案例进一步加深对数据挖掘技术的理解和认识。此外，本书在附录中，简单介绍了 Python 基本安装教程、线性代数、概率论及最优化理论的一些基础知识和概念，以帮助不同背景的读者查阅或复习相关知识点，从而更好地学习本书中介绍的数据挖掘技术。

　　本书适合作为理工类和经管类的高年级本科生或研究生的入门教材，以及对数据挖掘技术感兴趣的读者。本书采用英文编写，可作为高等学校数据挖掘双语课程建设的参考教材。通过本书的学习，读者能够为深入研究数据挖掘技术积累基础知识储备。

　　本书获得了南京财经大学教务处设立的全外语建设项目资助。感谢我的同事王月虎和吴士亮两位老师对书稿提出的宝贵意见，对提高本书质量很有帮助。感谢硕士研究生张珍和博士研究生姜李娟、桂冬冬在书稿相关资料收集、整理和编排过程中所作出的贡献。此外，本书是使用英文编写的数据挖掘入门教材，在写作过程中，参考了大量国外文献，为本书提供了丰富营养，使我受益匪浅。受时间和精力等因素影响，未能与论文和资料的作者一一取得联系，存在引用及理解不当之处敬请谅解，在此谨向他们致以崇高敬意和衷心的感谢！可以说没有这些重要的参考文献资料，完成本书将成为一个不可能的任务。

由于作者水平所限，再加上数据挖掘技术本身发展非常快，对一些新技术尚不够熟悉，因此，书中难免有错误和不妥之处，恳请各位专家和读者能够不吝赐教，给予批评和指正。

陈欣<br>
南京财经大学<br>
2022 年 1 月

资源索引

# Preface

Data mining, or knowledge discovery from data repositories, has emerged as one of the most exciting fields during the last decade, both in academic and industrial practice. Data mining aims at finding useful regularities in large datasets. Interests in this field are motivated not only by the growth of computerized data collections that are routinely kept by many organizations and commercial enterprises, but also by the high potential value of patterns discovered in those collections. For instance, bar code scanners at supermarkets produce extensive amounts of data about the commodities. An analysis of this type of data can reveal previously unknown, yet useful information about the shopping behavior of the customers. Data mining refers to a set of techniques that have been designed to efficiently discover interesting pieces of information or knowledge in huge datasets. Association rules, for example, are a class of hidden patterns that can be applied to identify which items tend to be purchased together by customers. There is currently a large commercial interest in the area, both for the development of data mining software and for the offering of consulting services on data mining.

Actually, data mining belongs to the interdisciplinary field and brings together many techniques from statistics, machine learning and information retrieval, etc. More importantly, lots of new techniques have recently emerged in many areas. Therefore it is very hard to give an exhaustive review of all techniques in a book, especially in a textbook for beginners. So my endeavor is to present the main data mining techniques currently applied including classification, clustering, association rule mining, artificial neural network and recent techniques for deep learning.

The target audience of this book could be either an advanced undergraduate or a first-year graduate or other professionals. It may also be used as a reference book for those who do not have much computer science background or are even beginners in this subject. Because each chapter is designed as independently as possible, you can focus on the topics that you are most interested in.

This book attempts to help understand the way you think about data and its role in businesses, governments, and even individuals creating massive collections of data as a by-product of their activity. It presents the principles and techniques of data mining, provides real-world examples and cases to put data-mining techniques in context to demonstrate how data analytics can be used to improve decision making. In particular, we will work "hands-on" with the data mining techniques by using Python/R packages.

The book is organized into several major parts: introduction, core techniques of data mining, data mining applications in management decision making, and appendix.

**The introduction part** outlines the knowledge required for basic ideas of data mining, such as the definition, implementing process and techniques, applications of decision making, and the development of data mining. This chapter attempts to offer readers the basic background knowledge of data mining concepts.

**The second part** introduces the core techniques of data mining, including classification, clustering, association rules, and artificial neural networks (deep learning, which is one of the most common algorithms presented in many other data mining courses). The corresponding contents cover usage explanation of the algorithm, pseudocode, examples and implementation codes with Python or R.

**The third part** draws attention to practical applications of data mining, which concentrate on several concrete instances concerning management decision making.

In the **appendix**, some basic prerequisites are reviewed to help readers to go over the related knowledge for a better understanding of the data mining techniques introduced in the book. The structure of this book is displayed in the following chart.

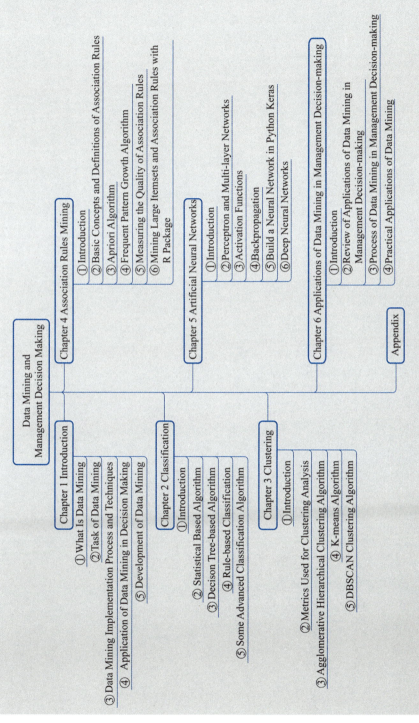

**The structure of this book**

# CONTENTS

**Chapter 1　Introduction** ·········· 1
   1.1　What Is Data Mining ·········· 2
   1.2　Tasks of Data Mining ·········· 3
   1.3　Data Mining Implementation Process and Techniques ·········· 8
   1.4　Application of Data Mining in Decision Making ·········· 13
   1.5　Development of Data Mining ·········· 16
   EXERCISES ·········· 20

**Chapter 2　Classification** ·········· 21
   2.1　Introduction ·········· 22
   2.2　Statistical Based Algorithm ·········· 26
   2.3　Decision Tree-based Algorithm ·········· 31
   2.4　Rule-based Classification ·········· 42
   2.5　Some Advanced Classification Algorithm ·········· 44
   EXERCISES ·········· 45

**Chapter 3　Clustering** ·········· 48
   3.1　Introduction ·········· 49
   3.2　Metrics Used for Clustering Analysis ·········· 51
   3.3　Agglomerative Hierarchical Clustering Algorithm ·········· 54
   3.4　K-means Algorithm ·········· 64
   3.5　DBSCAN Clustering Algorithm ·········· 70
   EXERCISES ·········· 75

**Chapter 4　Association Rules Mining** ·········· 77
   4.1　Introduction ·········· 78
   4.2　Basic Concepts and Definitions of Association Rules ·········· 79
   4.3　Apriori Algorithm ·········· 81
   4.4　Frequent Pattern Growth Algorithm ·········· 87
   4.5　Measuring the Quality of Association Rules ·········· 91
   4.6　Mining Large Itemsets and Association Rules with R Package ·········· 92
   EXERCISES ·········· 98

**Chapter 5　Artificial Neural Networks** ·········· 100
   5.1　Introduction ·········· 101
   5.2　Perceptron and Multi-layer Networks ·········· 103
   5.3　Activation Functions ·········· 107
   5.4　Backpropagation ·········· 111

  5.5 Build a Neural Network in Python Keras ⋯⋯ 124
  5.6 Deep Neural Networks ⋯⋯ 127
  EXERCISES ⋯⋯ 131

# Chapter 6 Applications of Data Mining in Management Decision-making ⋯⋯ 133

  6.1 Introduction ⋯⋯ 134
  6.2 Review of Applications of Data Mining in Management Decision-making ⋯⋯ 134
  6.3 Process of Data Mining in Management Decision-making ⋯⋯ 136
  6.4 Practical Applications of Data Mining ⋯⋯ 137
  EXERCISES ⋯⋯ 143

# Appendix ⋯⋯ 144

  A1 Basic Steps for Data Mining ⋯⋯ 144
  A2 A Quick Start Guide of Python for Beginners ⋯⋯ 146
  A3 Linear Algebra ⋯⋯ 163
  A4 Probability ⋯⋯ 168
  A5 Mathematical Optimization Basics ⋯⋯ 171

# References ⋯⋯ 177

# Chapter 1

# Introduction

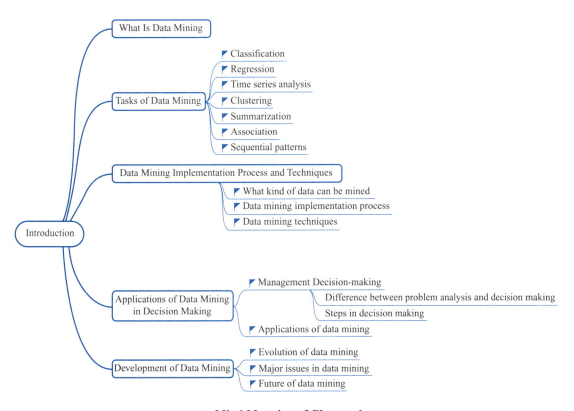

**Mind Mapping of Chapter 1**

　　随着数据量的急剧增长，如何高效处理这些数据，发现数据背后的有用信息，成为许多企业管理者在做出管理决策时的重要依据。特别是对于电子商务经营者来说，他们需要知道客户对过去购买产品的喜欢程度，可通过一些预测工具了解未来的购买情况等。从根本上说，数据挖掘是用于在庞大的数据量中寻找隐藏的、有效的和潜在的有用的模式和趋势，是用于发现数据之间未被怀疑或者以前未知的关系。通过数据挖掘获得的见解可用于营销、欺诈检测和科学发现等，帮助决策者做出决定或判断。根据 IBM 的分析，随着大数据的出现，数据挖掘技术的使用将更加普遍，这主要是由于信息量巨大，并且这些信息在性质和内容上具有多样化和广泛性。

数据挖掘任务通常可以分为两种类型：描述性任务和预测性任务。描述性任务检查数据的一般属性，而预测性任务是对可用数据集执行推理，以预测新数据集将如何响应。数据挖掘任务具体包括分类、预测、时间序列分析、关联、聚类、汇总等。所有这些任务都是预测性任务或描述性任务。例如医生根据患者的检测结果来诊断疾病类型可被视为一项预测性任务；零售商试图识别哪些商品被一起购买可被视为一项描述性任务等。因此，数据挖掘技术被广泛应用于医疗保健、保险、金融、零售、制造等行业。

数据挖掘技术给企业运营带来了巨大的变化。任何企业的决策过程都不能仅仅基于经验，尤其是在数据爆炸和信息广泛的竞争时代。当今，各个领域每秒都会产生大量数据，因此有必要了解可用于处理这些大量数据的各种工具。数据挖掘已成为所有领域标准业务实践的一部分，例如获取新客户、留住优质客户、增加现有客户的收入等。数据挖掘工具和软件的进一步发展也使企业管理者能够做出正确的决策，以实现利润最大化。

本章共分为五个小节。第一节介绍了数据挖掘概念，第二节分析了数据挖掘的任务，第三节讨论了数据挖掘实施过程和技术，第四节概述了数据挖掘在管理决策中的应用，第五节论述了数据挖掘的发展趋势。

## 1.1 What Is Data Mining

As the amount of data is growing dramatically, how to handle these data efficiently and uncover useful information and knowledge behind the data becomes an important task for many business managers in the process of making management decisions, especially for e-commerce operators, who are the main body of many new economies. For instance, online shop owners need to know how much customers like their products from past purchases, moreover, want to know future purchases by virtue of some prediction tools.

Fundamentally, data mining is used to look for hidden, valid, potentially useful patterns and trends in huge datasets. It is all about discovering unsuspected or previously unknown relationships amongst the data. The insights derived via data mining can be used for marketing, fraud detection, scientific discovery, etc., then you can decide or judge. Data mining can provide you potentially useful information that enables you to stand out from the crowd.

Data mining principles have been around for many years using machine learning, statistics, artificial intelligence (AI) and database technology. With the advent of big data, it is even more prevalent according to IBM. Big data caused an explosion in the use of more extensive data mining techniques, partially because the size of the information is much larger and because the information tends to be more varied and extensive in its very nature and content. Moreover, business-driven needs changed simple data retrieval and statistics into more complex data mining.

## 1.2 Tasks of Data Mining

Data mining tasks can be classified generally into two types of tasks, namely, descriptive and predictive. The descriptive tasks identify the general properties of data examined whereas predictive tasks perform inference on the available dataset to predict how a new dataset will respond. FIGURE 1-1 shows several data mining tasks such as classification, regression, time series analysis, prediction, clustering, association, summarization, sequential patterns etc., which are either predictive or descriptive tasks, and the list is illustrative rather than exhaustive. A medical practitioner trying to diagnose a disease based on the medical test results of a patient can be regarded as performing a predictive task. A retailer trying to identify products that are purchased together can be regarded as performing a descriptive task. These data mining techniques are applied in a wide range of industries, including healthcare, insurance, finance, retail, manufacturing, etc.

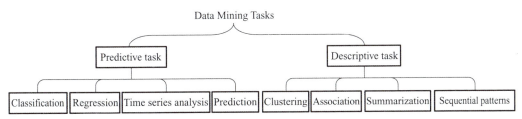

FIGURE 1-1   Data mining tasks

### 1.2.1 Classification

Classification is the task of finding models that analyze and classify a data item into several predefined classes. For example, you can easily classify cars into different types (sedan, 4×4, convertible) by identifying different attributes (number of seats, car shape, driven wheels). Given a new car, you might classify it into a particular class by comparing attributes with predefined classes. You can use the same principles for other applications, such as classifying customers by age, income, education into different groups. Additionally, you can use classification as a feeder for other techniques. For example, you can use decision trees to determine a classification, which can be applied to identify clusters by using common attributes in the different classifications. EXAMPLE 1-1 illustrates a simple example of classification.

 **EXAMPLE 1-1**

> A credit card company typically receives hundreds of thousands of applications for new cards. The application contains information regarding several different attributes, such as annual salary, any outstanding debts, age, etc. The problem is how to categorize applications for those who have good credit, those who have bad credit, and those who fall into a gray area (thus requiring further track record analysis).

### 1.2.2 Regression

Regression is the data mining task of identifying and analyzing the relationship between a dependent variable and one or more independent variables (also called 'predictors' 'covariates' or 'features').

Regression assumes that data fits into some type of function form, such as linear regression in which a researcher determines the line that most closely fits the given data by using some specific model selection criteria. Regression analysis is generally used for prediction and forecasting. In some situations, it can also be used to infer causal relationships between the independent and dependent variables. Importantly, one must carefully justify why existing relationships have predictive power for a new context or why a relationship between two variables has a causal interpretation. In EXAMPLE 1-2, a simple example of regression is shown.

#### EXAMPLE 1-2

> Suppose you're a sales manager trying to predict next month's revenue. You know that dozens, perhaps even hundreds of factors from the weather to a competitor's promotion to the rumor of a new and improved model can impact the revenue. In the case of multiple variable regression, you can find the relationship between temperature, pricing and the number of workers to the revenue. Thus, regression can analyze the impact of varied factors on business sales and profits.

### 1.2.3 Time series analysis

Time series analysis is a statistical technique that deals with time series data, or trend analysis, accounting for the fact that data points taken over time may have an internal structure (such as autocorrelation, trend, or seasonal variation). Time series data means that data is in a series of particular time periods or intervals (daily, weekly, hourly, etc.). This is opposed to cross sectional data which observes individuals, companies, etc., at a single point in time. In time series, time is often the independent variable and the goal is usually to make a forecast for the future. Three broad classes of time series models of prediction are the autoregressive (AR) models, the integrated (I) models, and the moving average (MA) models. Additionally, there are models, such as the autoregressive moving average (ARMA) and autoregressive integrated moving average (ARIMA) that are combinations of the above three. A time series plot (FIGURE 1-2) is used to visualize the time series. A time series example is given in EXAMPLE 1-3.

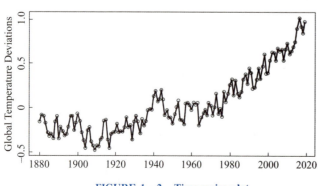

FIGURE 1-2  Time series plot

**EXAMPLE 1 – 3**

Consider the global temperature series record shown in Figure 1 - 2. The data is the global mean land-ocean temperature index from 1880—2020 with the base period of 1951—1980. In particular, data deviations are measured in degrees centigrade, from the 1951—1980 average. We note an apparent upward trend in the series during the latter part of the twentieth century that has been used as an argument for the global warming hypothesis. The question of trend is of particular interest to global warming proponents and opponents which can be used as evidence to show whether the overall trend is natural or whether it is caused by some human induced interface.

### 1.2.4 Clustering

Clustering is similar to classification except that the groups are not predefined instead of being defined by the data itself. It is alternatively referred to as unsupervised learning or segmentation. Clustering analysis is the main task of discovering groups and clusters in the data in such a way that objects in the same group (called a cluster) are more similar (in some sense) to each other than to those in other groups (clusters). By examining one or more attributes or classes, you can group individual pieces of data to form a structure opinion. At a simple level, clustering is using one or more attributes as your basis for identifying a cluster of correlating results. Clustering is useful to identify different information because it correlates with other examples so you can see where the similarities and ranges agree. It is widely used in market research when working with multivariate data from surveys and test panels. Based on the attributes of the cluster, market researchers use clustering analysis to partition the general population of consumers into market segments and to better understand the relationships between different groups of consumers/potential customers, and for use in market segmentation, product positioning, new product development and selecting test markets. EXAMPLE 1 – 4 and FIGURE 1 – 3 provide a good illustration of clustering.

It should be noted that segmentation is often considered to be identical to clustering. In some applications, segmentation is viewed as a specific type of clustering.

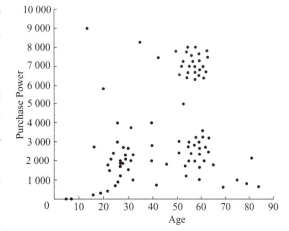

**FIGURE 1 – 3   Clustering**

In our text, we refer to two expressions, clustering and segmentation, interchangeably.

EXAMPLE 1-4

A sample of sales data compares the age of the customer to the sale size. Generally, people can have more disposable income in their sixties (when the children have left home) than that in their twenties (before marriage and kids) and fifties. We can identify two clusters: one around the US $ 2 000/20 - 30 age group and another at the US $ 7 000 - 8 000/50 - 65 age group. In this case, we've both hypothesized and proved our hypothesis with a simple graph that we can create using any suitable graphing software for a quick manual view. More complex determinations can be achieved by various algorithms that differ significantly in their understanding of what constitutes a cluster and how to efficiently find them.

### 1.2.5 Summarization

Summarization is one of the key data mining concepts which involve techniques for finding a compact description of a dataset. Simple summarization methods such as tabulating the mean and standard deviations are often applied for exploratory data analysis, data visualization and automated report generation. Summarization is also called *characterization* or *generalization* and can be viewed as compressing a given set of transactions into a smaller set of patterns while retaining the maximum representative information. EXAMPLE 1-5 illustrates the process.

 EXAMPLE 1-5

A customer relationship manager in *All-Electronics* may order the following data mining task: Summarize the characteristics of customers who spend more than $ 5000 a year at *All-Electronics*. The result is a general profile of these customers, such as that they are 40 to 50 years old, employed, and have excellent credit ratings. The data mining system should allow the customer relationship manager to drill down on any dimension, such as on occupation to view these customers according to their type of employment. The data can be summarized in different abstraction levels and from different angles.

### 1.2.6 Association

Association refers to the data mining task of uncovering the connection among a set of items. Association analysis is used for commodity management, advertising, catalog design, direct marketing, etc. A retailer can identify the products that normally customers purchase together or even find the customers who respond to the promotion of the same kind of products. If a retailer finds that *beer and diapers* are purchased together mostly, he can put nappies together with the sale of beer. Note here, association rules do not represent any causal relationship inherent in the actual data or the real world. EXAMPLE 1-6 illustrates the use of association rules in market analysis.

 **EXAMPLE 1-6**

> Imagine that you are a sales manager at *All-Electronics*, and you are talking to a customer who recently bought *a PC and a digital camera* from the store. What should you recommend to her next? Information about which products are frequently purchased by your customers following their purchases of a PC in sequence would be very helpful in making your recommendations. You can then use the results to plan marketing or advertising strategies, or in the design of a new catalog. Items that are frequently purchased together can be placed in proximity to further encourage the combined sale of such items. If customers who purchase computers also tend to buy antivirus software at the same time, then placing the hardware display close to the software display may help increase the sales of both items.

### 1.2.7 Sequential patterns

Sequential patterns are a useful method for identifying trends, or regular occurrences of similar events based on a time sequence of actions which are similar to associations in that data are found to be related based on time. For example, a pattern of buying a PC first, then a digital camera, and then a memory card, occurs frequently in the shopping record database, then it is a sequential (frequent) pattern.

It's particularly useful for data mining transactional data. With customer data you can identify that customers buy a particular collection of products together at different times of the year. Understanding sequential patterns can help organizations recommend additional items to customers to spur sales. For instance, this technique can reveal what items of clothing customers are more likely to buy after an initial purchase of a pair of shoes. In a shopping basket application, you can use this information to automatically suggest that certain items be added to a basket based on their frequency and past purchasing history. EXAMPLE 1-7 presents the discovery of some simple patterns.

 **EXAMPLE 1-7**

> The webmaster at the XYZ Corp. periodically analyzes the weblog data to determine how users of the XYZ web pages access them. He is interested in determining what sequences of pages are frequently accessed. He determines that 75% of the users of page A follow one of the following patterns of behavior: $<A, B, C>$ or $<A, D, B, C>$ or $<A, E, B, C>$. He then determines to add a link directly from page A to page C.

## 1.3 Data Mining Implementation Process and Techniques

### 1.3.1 What kind of data can be mined

Knowing the type of data that you're trying to mine is the first critical step to implement data mining applied to obtain the best results. In principle, data mining is not specific to one type of media or data. Data mining should apply to any kind of information repository. However, algorithms and approaches may differ when applied to different types of data. Indeed, the challenges presented by different types of data vary significantly. Data mining is being put into use and studied for databases, including relational databases, object-relational databases, object-oriented databases, data warehouses, transactional databases, unstructured and semistructured repositories such as the World Wide Web (WWW), advanced databases such as spatial databases, multimedia databases, time series databases, textual databases, and even flat files.

### 1.3.2 Data mining implementation process

Generally, there are 6 steps of the data mining standard process which include business understanding, data understanding, data preparation, modeling, evaluation, and deployment, data mining process below in FIGURE 1-4. It is also known as the cross-industry standard process for data mining (CRISP-DM), conceived in 1996 after going through many workshops, and making contributions to more than 300 organizations. Notably, this is an open standard process model that describes common approaches used by data mining experts.

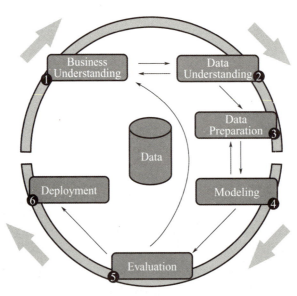

**FIGURE 1-4   Data mining process**

(1) Business understanding

In this phase, business and data mining goals are established.

- First, you need to understand business objectives clearly and find out what your client wants (even they do not know themselves).
- Next, assess the current data mining scenario. Factor in resources, assumptions, constraints, and other significant factors should be considered in your assessment.
- Then, define your data mining goals from the business objectives and current situations.
- Finally, a good data mining plan is very detailed and should be established to achieve both business and data mining goals. The plan should be as detailed as possible.

(2) Data understanding

In this phase, a sanity check on data is performed to make sure whether it is appropriate for the data mining goals built in the above step.

- This phase starts with data collection from multiple and available data sources. Some issues like object matching and schema integration can arise during the data integration process. For example, table A contains an entity named *cust_no* whereas table B contains an entity named *cust-id*.
- Next, the step is to examine the properties of the acquired data carefully.
- Then, the data is required to be explored by answering the data mining questions (decided in the business understanding phase), which can be addressed by employing the query, reporting and visualization tools.
- Finally, based on the results of the query, the data quality should be ascertained, such as completeness, validity and integrity.

(3) Data preparation

In this phase, data is prepared readily for final use in modeling. It is about 90% of the total project time to prepare the final dataset. The identified data is required to be selected, cleaned, transformed, formatted, anonymized, and formatted into the desired form. Data transformation operations could contribute toward the success of the data preparation process, such as smoothing, aggregation, generalization, normalization, attribute construction, etc.

(4) Modeling

In this phase, various modeling methods are selected and applied. The parameters are measured to optimum values. Some techniques need particular requirements in the form of data. Therefore, stepping back to the data preparation phase is necessary.

- Firstly, according to the business objectives, suitable modeling techniques should be selected for the prepared dataset.
- Next, one should generate a procedure or mechanism for testing the validity and quality of the model before constructing a model. Typically you can separate the dataset into training and test set, build the model on the training set and assess its

quality on the separate test set.
- Then, you can run the model on the prepared dataset.
- Results should be assessed by all stakeholders to make sure that the model can meet data mining objectives based on domain expertise, data mining success criteria and the required design, etc.

(5) Evaluation

In this phase, patterns identified are evaluated in the context of the business objectives.
- Firstly, results generated by the data mining model should be evaluated against the business objectives.
- Secondly, obtaining business understanding is an iterative process. While understanding and discovering the model results, new business requirements may be raised due to data mining.
- Finally, a go or no-go decision must be taken to move the model to the deployment phase.

(6) Deployment

Deployment refers to how the outcomes need to be applied to everyday business operations. The knowledge or information discovered during data mining process should be organized and presented in an easy way to understand for non-technical stakeholders and clients. Detailed deployment plans for maintenance, and monitoring of data mining discoveries have to be created for implementation and future support. During the project A final project report is drawn up with lessons learned and key experiences from the project viewpoint. Reviewing what went right, what was done wrong, then it can be used to improve the organization's performance.

### 1.3.3　Data mining techniques

As a multi-disciplinary field, data mining integrates many approaches and techniques from various disciplines such as machine learning, statistics, artificial intelligence, neural networks, database management, data warehousing, data visualization, spatial data analysis, probability graph theory, etc. Some basic and important techniques of data mining are listed below.

(1) Statistical techniques

Statistical techniques are at the core of most analytics involved in data mining tasks. They are also the basis for many other data mining techniques. Two main statistical methods are used in data analysis: descriptive statistics, which summarize data from a sample using indexes such as the mean or standard deviation, and inferential statistics. Statistics is the traditional field that deals with the quantification, collection, analysis, interpretation, and conclusions from data. Point estimation, bayes theorem, hypothesis testing, and correlation are common statistical concepts widely used in the data mining area. Note that applying statistical methods in data mining is far from trivial. A serious challenge is how to apply a statistical method to a large dataset, typically the size is from $10^7$ to $10^{10}$ Bytes. Furthermore, one needs to distinguish between the number of cases (observations) in a large dataset ($n$), and the number of features (variables) available for each case ($m$). In a large dataset, $n$, $m$ or both can be large.

(2) Decision trees

A decision tree is a specific type of predictive model that lets business and industrial organizations effectively mining data. Technically, a decision tree is a part of machine learning covering both classification and regression and is popular for its extremely straight forward nature. For decision analysis, a decision tree can be used to visually and explicitly represent decisions and decision making, which enables users to clearly understand how the data inputs affect the outputs. In general, decision tree algorithms are referred to as classification and regression trees (CART). Decision tree example see FIGURE 1 – 5.

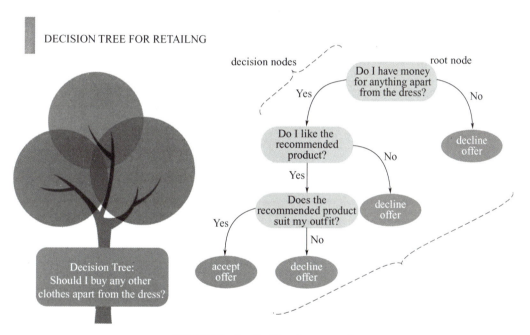

FIGURE 1 – 5  Decision tree example

(3) Visualization

Data visualization is the graphical representation of the data and information extracted from data mining. By using visual elements like graphs, charts, and maps, data visualization tools and techniques can help in analyzing massive amounts of information and making data-driven decisions. They grant users insights into data based on sensory perceptions that people can remember and memorize large chunks of data at a single glance. Today's data visualizations are dynamic, useful for streaming data in real-time, and characterized by different colors that provide an accessible way to see and understand trends, outliers, and patterns in data. Dashboards are a powerful way to use data visualizations to uncover data mining insights. It has been used by various organizations with different metrics to visually highlight patterns in data, instead of simply using numerical outputs of statistical models. CEO dashboard see FIGURE 1-6.

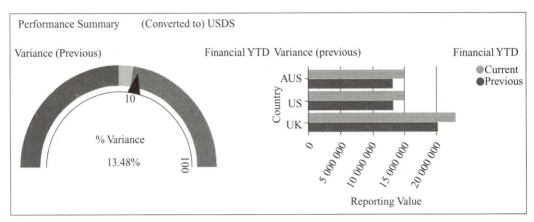

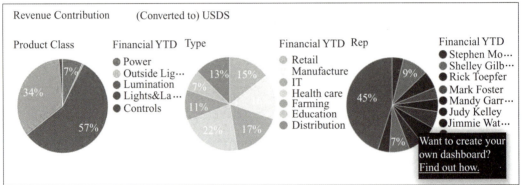

**FIGURE 1-6　CEO dashboard**

(4) Data warehousing

A data warehouse is a large collection of business data used to assist enterprises to make decisions. The data stored in data warehouses come from different sources such as internal applications including marketing, sales, and finance, customer-facing apps, and external partner systems. From a technical viewpoint, a data warehouse periodically extracts data from those apps and systems, then the data passes through formatting and import processes to match the data already in the warehouse. The data warehouse stores this processed data for decision-makers to use. Data warehouses provide online analytical processing (OLAP) tools for the interactive analysis of multidimensional data of varied granularities, which facilitates effective data generalization and data mining. Classically, data warehousing involved storing structured data in relational database management systems. However, as businesses make the move to the cloud, so do the databases and data warehousing tools. There are many cloud data warehouses and data warehouses in semi-structured and unstructured data stores like *Hadoop*. The cloud offers many advantages, such as flexibility, collaboration, and accessibility from anywhere. Although data warehouses were traditionally used for historical data, many modern approaches can provide an in-depth, real-time analysis of data.

(5) Machine learning and artificial intelligence

Machine learning and AI represent the most advanced developments in data mining. Machine learning is viewed as a subset of AI. Performing machine learning involves creating a model, which is trained on some training data and then can process additional data to make predictions. Various advanced types of models have been researched and developed for machine learning systems, such as deep learning, reinforcement learning, which can offer highly accurate predictions when working with data at a large scale. They're useful for processing data in AI deployments like computer vision, speech recognition, or sophisticated text analytics using Natural Language Processing (NLP). These data mining techniques are also adaptable for determining value from semi-structured and unstructured data.

(6) Web mining

Web mining allows the searching for patterns in data through *content mining, structure mining, and usage mining*. Content mining is used to examine data collected by search engines. Structure mining is used to examine data related to the structure of a particular web site. Usage mining is used to examine data related to a particular user's browser as well as data gathered by forms that the user may have submitted during web transactions. Web mining, as a sequential pattern mining application, is concerned with finding user navigational patterns on the World Wide Web by extracting knowledge from weblogs. An example of applying sequential pattern mining in this topic would be searching for a pattern in sites visited. Given websites $a$ thru $f$, if a log pattern showed a frequent sequence, *abac*, for instance, you would see that a user that visits $a$ tends to visit $b$ and then re-visit $a$ before continuing to $c$.

## 1.4 Application of Data Mining in Decision Making

### 1.4.1 Management decision-making

It has been shown that managers can be required to make hundreds of decisions in the course of an average week at work. Yet some decisions are so complex that a single decision can take weeks, months, or even longer to be final. Whatever the number or complexity of decisions in your workplace, it's important to know decision making is a core management competency for managers to have opportunities to review their approach and make adjustments if necessary. And effective management decision making often involves the selection and application of appropriate tools and having awareness of where each can be useful.

(1) Difference between problem analysis and decision making

While they are related, problem analysis and decision making are distinct activities. Decisions are commonly focused on a problem or challenge. Decision makers must gather and consider data before making a choice. Problem analysis involves framing the issue by defining its boundaries, establishing criteria with which to select from alternatives, and developing conclusions based on available information. Analyzing a problem may not result in a decision, although the results are an important ingredient in all decision making.

(2) Steps in decision making

Management decisions can be viewed to have two key components: *content* and *process*. Content refers to data information and knowledge on which a decision is based whereas decision-making process refers to the steps you go through to make a decision. Whilst the content component is unique to every decision, the following process steps should be involved in every decision regardless of its simplicity or complexity.

**Step 1:** Defining the issue or problem for which a decision is needed.

**Step 2:** Identifying who has the authority to make the decision.

**Step 3:** Recognizing who should be consulted in the decision making and why.

**Step 4:** Gathering the right decision making content (data information and knowledge) from appropriate sources including the place of consultation.

**Step 5:** Analyzing the content and generating options by using data mining tools.

**Step 6:** Critically evaluating the alternatives.

**Step 7:** Making the decision by choosing the best option.

**Step 8:** Communicating the decision.

**Step 9:** Implementing the decision and reviewing the effects of the decision.

All too often especially when managers are under stress, one or more of these steps is omitted or compromised. Then what could have been an effective decision turns into a decision with adverse consequences. For example, taking the time to properly define the problem or issue (**Step 1**) provides a solid foundation for the following steps. When managers replace the symptoms for a proper problem definition, they adventure gathering and acting on the wrong data. And when managers avoid consulting others who may be affected by the decision (**Step 3**), they risk alienating people who might otherwise have been committed to the decision.

### 1.4.2 Applications of data mining

Data mining helps businesses identify important facts, trends, patterns, relationships and exceptions that are normally unnoticed or hidden. Today data mining is widely used in diverse areas based on its importance at the heart of analytic efforts across a variety of industries and disciplines. A comprehensive list of data mining application fields and the usage is given in TABLE 1-1. By analyzing the purchase patterns of customers, retailers can come up with smarter marketing promotions and campaigns which will, in turn, increase the sales. With market segmentation, retailers can identify customers who purchase the same products. So, they can come up with new products at the right time by analyzing the interests and demographics of customers. Data mining can also be used to predict customers' demand who are most likely to transfer purchasing from these market competitors. Fraud detection is a major headache for finance and insurance companies. Studies show that customer demographics can be effectively used to predict their fraudulent nature. Nowadays, data mining is used to identify transactions that are most likely to be fraudulent. In the healthcare industry, data mining techniques are mainly used for most accurate disease diagnosis and most effective treatments. It is also helpful in predicting health insurance fraud, healthcare cost and length of stay (LOS) of hospitalization.

TABLE 1-1  A comprehensive list of data mining application fields and the usage

| Applications | Usage |
| --- | --- |
| Communication | Data mining techniques are used in the communication sector to predict customer behavior to offer highly targeted and relevant campaigns |
| Insurance | Data mining helps insurance companies to price their products reasonably and promote new offers to their new or existing customers |
| Education | Data mining benefits educators to access student data, predict achievement levels and find students or groups of students who need extra attention. For example, students who are weak in math subject |
| Manufacturing | With the help of data mining manufacturers can predict the wear and tear of production assets. They can anticipate maintenance which helps them minimize downtime |
| Banking | Data mining helps the finance sector to get a view of market risks and manage regulatory compliance. It helps banks to identify probable defaulters to decide whether to issue credit cards, loans, etc |
| Retail | Data mining techniques help retail malls and grocery stores identify and arrange the most sellable items in the most attentive positions. It helps store owners to come up with an offer which encourages customers to increase their spending |
| Service Providers | Service providers like mobile phone and utility industries use data mining to predict the reasons when a customer leaves their company. They analyze billing details, customer service interactions, complaints made to the company to assign each customer a probability score and offer incentives |
| E-commerce | E-commerce websites use data mining to offer cross-sells and up-sells through their websites. One of the most famous names is Amazon, who uses data mining techniques to get more customers into its E-commerce store |

| Applications | Usage |
| --- | --- |
| **SuperMarkets** | Data mining allows supermarkets to develop rules to predict if their shoppers were likely to be expecting. By evaluating their buying pattern, they could find woman customers who are most likely pregnant. They can start targeting products like baby powder, baby shops, diapers, and so on |
| **Crime Investigation** | Data mining helps crime investigation agencies to deploy police workforces (where is a crime most likely to happen and when), who to search at a border crossing, etc |
| **Bioinformatics** | Data mining helps to mine biological data from massive datasets gathered in biology and medicine |

## 1.5  Development of Data Mining

### 1.5.1  Evolution of data mining

The term "data mining" has evolved over a long period of time. Early methods of identifying patterns in data include Bayes' theorem and regression analysis. Data mining has evolved into a mainstream technology because of two complementary, yet antagonistic phenomena: (a) the data deluge, fueled by the maturing of database technology and the development of advanced automated data collection tools; (b) the starvation for knowledge, defined as the need to filter and interpret all these massive data volumes stored in databases, data warehouses and other information repositories. As datasets have grown in size and complexity, direct hands-on data analysis has increasingly been augmented with indirect, automatic data processing. Date Ming (DM) can be thought of as the logical succession to Information Technology (IT). The development of Data Warehouse (DW) and Decision Support System (DSS) in the early 1990s allowed the manipulation of data coming from heterogeneous sources and supported multiple-level dynamic and summarized data analysis. After that, data mining terms began to be known and applied within various research communities by statisticians, data analysts, and the Management Information System (MIS) communities. The development of data mining is shown in FIGURE 1-7.

The following are some major milestones in the past 10 years, which indicate that data mining is being evolved and blended with data science and big data.

- In 2009, the term NoSQL was reintroduced (a variation had been used since 1998) by Johan Oskarsson, when he organized a discussion on "open-source, non-relational databases".

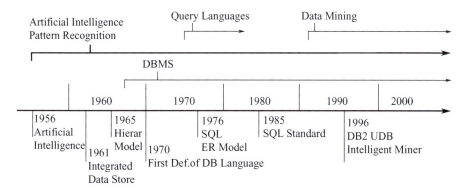

**FIGURE 1-7　The development of data mining**

- In 2011, job listings for data scientists increased by 15000%. There was also an increase in seminars and conferences devoted specifically to data science and big data. data mining had proven itself to be a source of profits and had become a part of the corporate culture.
- In 2011, James Dixon, CTO of Pentaho promoted the concept of Data Lakes, rather than Data Warehouses. Dixon stated the difference between a Data Warehouse and a Data Lake is that the Data Warehouse categorizes the data at the point of entry, wasting time and energy, while a Data Lake accepts the information using a non-relational database (NoSQL) and does not categorize the data, but simply stores it.
- In 2013, IBM shared statistics showing 90% of the data in the world had been created within the last two years.
- In 2015, using deep learning techniques, google's speech recognition, google Voice, experienced a dramatic performance jump of 49%.
- In 2015, Bloomberg's Jack Clark, wrote that it had been a landmark year for AI. Within google, the total of software projects using AI increased from "sporadic usage" to more than 2700 projects over the year.
- In February 2015, DJ Patil became the first Chief Data Scientist at the White House.
- Present, one of the most active techniques being explored today is *Deep Learning*. Capable of capturing dependencies and complex patterns far beyond other techniques, it is reigniting some of the biggest challenges in the world of data mining, data science and artificial intelligence.

Today, during its evolution, data mining has become an important part of business and academic research and is widespread in business, science, engineering and medicine. Technically, this includes machine translation, robotics, speech recognition, the digital economy, and search engines. Mining of credit card transactions, stock market movements, national security, genome sequencing and clinical trials are just the tip of the iceberg for data mining applications. Terms like big data are now commonplace with the collection of data becoming cheaper and the proliferation of devices capable of collecting data. Data science now influences economics, governments, business and finance.

### 1.5.2　Major issues in Data mining

Data mining embodies techniques that have sometimes existed for many years but have only lately been applied as reliable and scalable tools and again outperform older classical statistical methods. However, before data mining develops into a conventional, mature and trusted discipline, many still pending issues have to be addressed. Some of these issues are addressed below. Note that these issues are not exclusive and are not ordered in any way.

**Security and social issues:** Security is an important issue with any data collection that is shared and/or is intended to be used for strategic decision-making. In addition, when data is collected for customer profiling, user behavior understanding, correlating personal data with other information, etc., large amounts of sensitive and private information about individuals or organizations is collected and stored. This becomes controversial given the confidential nature of this data and the potential illegal access to the information. Moreover, data mining could disclose new implicit knowledge about individuals or groups that could be against privacy policies, especially if there is potential dissemination of discovered information. Another issue that arises from this concern is the appropriate use of data mining. Due to the value of data, databases of all sorts of content are regularly sold, and because of the competitive advantage that can be attained from implicit knowledge discovered, some important information could be withheld, while other information could be widely distributed and used without control.

**User interface issues:** The knowledge discovered by data mining tools is useful as long as it is interesting, and above all understandable by the user. Good data visualization eases the interpretation of data mining results, as well as helps users better understand their needs. Many data exploratory analysis tasks are significantly facilitated by the ability to see data in an appropriate visual presentation. There are many visualization ideas and proposals for effective data graphical presentation. However, there is still much research to accomplish in order to obtain good visualization tools for large datasets that could be used to display and manipulate mined knowledge. The major issues related to user interfaces and visualization are "screen real-estate", information rendering, and interaction. Interactivity with the data and data mining results is important since it provides ways for customers to focus and refine mining tasks, as well as to depict the discovered knowledge from different perspectives and at different conceptual levels.

**Mining methodology issues:** These issues pertain to the data mining approaches applied and their limitations. Topics, such as versatility of the mining approaches, the diversity of data available, the dimensionality of the domain, the broad analysis demands (when known), the evaluation of the knowledge discovered, the exploitation of background knowledge and metadata, the control and handling of noise in data, etc., are all issues that can affect mining techniques choices. For instance, it is often expected to have different data mining methods available since different approaches may perform differently depending upon the data at hand. Moreover, different approaches may be suitable and meet clients' demands differently. Furthermore, most algorithms assume data to be *noise-free*, which is obviously a *strong assumption*.

Many datasets contain exceptions, invalid or incomplete information, or outlier data, etc., which may intricate, if not obscure, the analysis process and in many cases compromise the accuracy of the results. Consequently, data preprocessing (data cleaning and transformation) becomes very critical. Although data cleaning is often viewed as time-consuming and frustrating, it is one of the most important phases in data mining and knowledge discovery process. Data mining techniques should be able to handle noises in data or incomplete information.

Notably, rather than data size, the size of the searching space is even more decisive for data mining techniques, which often depends upon the number of dimensions in the domain space. It usually grows exponentially when the amount of dimensions increases. This is known as the curse of dimensionality. This "curse" affects so badly the performance of some data mining approaches that it is becoming one of the most urgent issues to solve.

**Overfitting and underfitting:** When a model is generated that is associated with the given dataset, it is desirable that the model also fits other database sets. Usually, a learning algorithm is trained using the set of "training data" for which the desired output is known. The goal is that the algorithm will also perform well on predicting the output when fed "validation data" that was not encountered during its training. *Overfitting* happens when a model learns the detail and noise in the training data to the extent that it negatively impacts the performance of the model on new data. This means that the noise or random fluctuations in the training data is picked up and learned as concepts by the model. The problem is that these concepts do not apply to new data and negatively impact the model's ability to generalize. *Underfitting* occurs when a statistical model or machine learning algorithm cannot adequately capture the underlying structure of the data. It occurs when the model or algorithm does not fit the data enough. *Underfitting* occurs if the model or algorithm shows low variance but high bias (to contrast the opposite, overfitting from high variance and low bias). It is often a result of an excessively simple model. In reality, the data often studied has some degree of error or random noise within it. Thus, attempting to make the model conform too closely to slightly inaccurate data can infect the model with substantial error and reduce its predictive power of generalization.

### 1.5.3 Future of data mining

Data mining techniques have brought about tremendous changes in business operations. A huge amount of data is generated every second in various fields and it is necessary to have knowledge of various tools that can be employed to process this huge data. Decisions making process in any organization or business cannot be based on experience alone, especially now in the age of data explosion and a wide range of information and competition. Data mining has become part of standard business practice in many areas such as acquiring new customers, retaining good customers, increasing revenue from existing customers. The further development of data mining tools and software has also made it possible for business owners to make the right decisions to maximize their profits. Though data mining has got its challenges, we should have confidence that data mining tools and techniques would be significantly improved based on futuristic challenges that might occur due to data complications.

## EXERCISES

1. Identify and elaborate on what data mining is.
2. Describe the data mining implement process.
3. How many data mining techniques can you present besides those mentioned in section 1.3?
4. Find some examples of data mining applications that have been used in management decision making?
5. What is your personal viewpoint on the future of data mining?

本章配套资源

# Chapter 2

# Classification

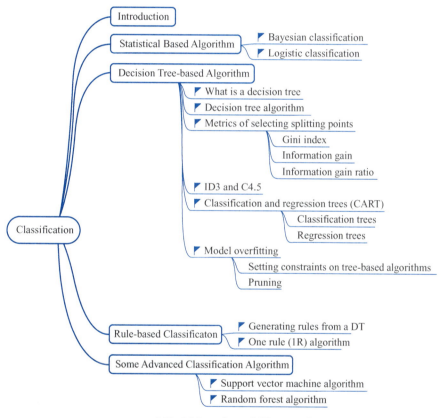

**Mind Mapping of Chapter 2**

分类是根据数据的相似性或共同标准将给定数据集分成不同类别的过程。它帮助人类分析存在于他们周围的事物、对象和想法，并简化他们对世界的理解。一般来说，分类可以分为二元分类和多元分类两种。因此，数据集可能只是二元或多元。这些类通常被称为目标、标签或类别。请注意：即使是连续的，估计和预测也可以被认为是分类。有许多分类应用示例，如图像和模式识别、贷款审批、医疗诊断、生物特征识别、文档分类和金融市场趋势分类等。本章的例 2-1 是 MNIST 手写数字识别和分类问题，其目的是将每张输入照片（矢量图）分配给有限离散数（0-9）中之一。MNIST 数据库包含 60000 张训练图像和 10000 张测试图像，由 Yann LeCun、Corinna Cortes 和 Christopher Burges 使用美国国家标准与技术研究院提供的大量扫描文档数据集开发，这也是数据集名称的来源，即 MNIST 数据集。每个图像都是一个 28×28 像素的正方形（总共 784 像素）。

给定数据库 $D=\{t_1, t_2, \ldots, t_n\}$ 和分类集 $C=\{C_1, C_2, \ldots, C_m\}$，则分类问题是定义一个映射函数 $f: D \to C$，其中每个 $t_i$ 都分配给一个类 $C_i$，即 $C_j\{t_i | f(t_i)\}=C_j$，$1 \leqslant i \leqslant n$ 且 $t_i \in D$。请注意：预定义的类不应重叠，分类函数有助于从数据库映射到类集，数据库中的每个元组都被分配给一个类。机器学习中有许多分类算法，例如逻辑回归、朴素贝叶斯、决策树、随机森林等。实现分类的算法，特别是在具体实现中被称为分类器。术语"分类器"有时也指数学函数，由将输入数据映射到类别或类的分类算法中实现，它是基于距离的算法采用相似度或距离度量来执行分类任务；统计算法具有基于统计信息特征；决策树和神经网络方法使用结构特征来执行分类；基于规则的分类算法生成 if-then 规则来执行分类。本章重点介绍基于统计的、基于决策树的和基于规则的分类方法，其余的算法将在其他章节中介绍。

## 2.1 Introduction

**Classification** is a process of categorizing a given set of data into categories (classes, types, index) based on their similarities or common criteria. It allows humans to organize things, objects, and ideas that exist around them and simplify their understanding of the world. Generally, classification can be broken down into two areas: binary classification (like identifying whether the person is male or female or that the mail is spam or not) and multi-class classification. Accordingly, the dataset may simply be bi-class or multi-class. The classes are often referred to as target, label, or categories. Note that estimation and prediction can also be thought of as classification even for continuity. There are many examples of classification applications including image and pattern recognition (see EXAMPLE 2-1), loan approval, medical diagnosis, biometric identification, document classification, and classifying financial market trends.

 **EXAMPLE 2-1**

> A popular example of a classification problem would be MNIST handwritten digit recognition, in which the purpose is to assign each input photo (vectors) to one of a finite discrete number (0-9). The MNIST database contains 60 000 training images and 10 000 testing images, developed by Yann LeCun, Corinna Cortes and Christopher Burges by using a number of scanned document dataset available from the National Institute of Standards and Technology (NIST). This is also where the name for the dataset comes from, as the Modified NIST or MNIST dataset. Each image is a 28×28 pixel square (784 pixels total) as is shown in the following pictures (FIGURE 2-1).

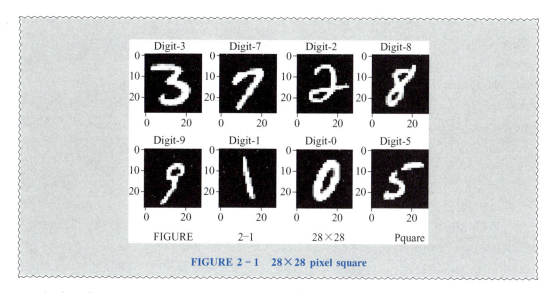

**FIGURE 2 – 1    28×28 pixel square**

A classification problem can be defined as: Given a database $D=\{t_1, t_2, \ldots, t_n\}$ of tuples (namely items or records) and a set of classes $C=\{C_1, C_2, \ldots, C_m\}$, the **classification problem** is to define a mapping function $f: D \to C$ where each $t_i$ is assigned to one class $C_j$ containing precisely those tuples mapped to it; that is, $C_j = \{t_i \mid f(t_i)\} = C_j\ 1 \leqslant i \leqslant n$, and $t_i \in D$. Note that the predefined classes should not be non-overlapping, and the classification function help to map from the database to the set of classes. Each tuple in the database is assigned to exactly one class.

There are numerous algorithms for classification in machine learning like logistic regression, naive Bayes, decision trees, random forests, and many more. An algorithm that implements classification, especially in a concrete implementation, is known as a classifier. The term "classifier" sometimes also refers to the mathematical function, implemented by a classification algorithm that maps input data to a category or class. Classification algorithms can be represented according to the categorization as shown in FIGURE 2 – 2. Distance-based algorithms employ similarity or distance measures to perform the classification task. Statistical algorithms are based directly on the use of statistical information. Decision tree and NN approaches use these structures to perform the classification. Rule-based classification algorithms generate *if-then* rules to perform the classification.

In the following sections, we only examine the three algorithms including the statistical-based, decision tree-based, and rule-based classification methods. As the rest algorithms would be covered in other chapters.

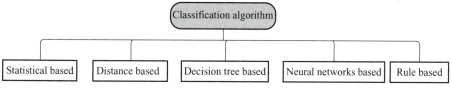

**FIGURE 2 – 2    Categorization of Classification algorithms**

As to the performance measurement of the classification algorithms, it is usually examined by evaluating the accuracy of the classification. In addition, space and time rules can also be used to determine the choice of the appropriate classification algorithms but only for a secondary consideration. The accuracy of classification is usually obtained by calculating the percentage of the number of correctly classified patterns to the total number of patterns. It can also be defined as the ratio of the sum of true positives (TP) and true negatives (TN) to the total number of training items [sum of TP, false positives (FP), false negatives (FN) and TN].

$$\text{Classification accuracy(ACC)} = \frac{TP+TN}{TP+FP+FN+TN}$$

However, classification accuracy alone can be misleading if you have an unequal number of observations in each class or if you have more than two classes in your dataset. In this situation, a confusion matrix is a useful tool for summarizing the performance of a classification algorithm. Calculating a confusion matrix can give you a better idea of what your classification algorithm is getting right and what types of errors it is making. The following example is from *Wikipedia*, suppose we have $P$ positive instances (the number of real positive cases in the data) and $N$ negative instances (the number of real negative cases in the data) from some experiments. The four outcomes can be formulated in a $2 \times 2$ contingency table or confusion matrix in TABLE 2-1.

TABLE 2-1　Confusion matrix(sometimes contingency table)

|  |  | True condition | |
|---|---|---|---|
|  |  | **Condition positive** | **Condition negative** |
| Predicted condition | Positive | True positive, TP | False positive, FP<br>Type *one* error |
|  | Negative | False negative, FN<br>Type *two* error | True negative, TN |
|  | Total population | $P$=(TP+FN) | $N$=(FP+TN) |

Particularly, the True positive rate (TPR) = $\frac{TP}{TP+FN} = \frac{TP}{P}$, and false positive rate (FPR) = $\frac{FP}{TP+TN} = \frac{FP}{N}$. The TPR defines how many correct positive results occur among all positive samples available during the training, and is also known as *recall*. FPR, on the other hand, defines how many wrong positive results occur among all negative samples available during the training. To draw a receiver operating characteristic curve (ROC curve), only TPR and FPR are required (as functions of some classifier parameter) to serve as horizontal and vertical axes, respectively, which depicts relative trade-offs between true positive (benefits) and false positive (costs).

Each prediction result or instance of a confusion matrix represents one point in the ROC space derived from varied discrimination thresholds. The best possible classification algorithm would produce a point in the upper left corner or coordinate (0,1) of the ROC space, representing 100%

sensitivity (no false negatives) and 100% specificity (no false positives). The (0,1) point is also called a perfect classification. A random guess would give a point along a diagonal line (the so-called line of no-discrimination) from the left bottom to the top right corners (regardless of the positive and negative base rates) which divides the ROC space into two equal parts. Points above the diagonal represent good classification results (better than random) while the points below the line represent bad results (worse than random). Moreover, the closer a point is to the upper left corner, the better it predicts. TABLE 2-2 gives four concrete examples of prediction results with 100 positive and 100 negative instances.

TABLE 2 – 2  Four concrete examples

| A | | | B | | | C | | | C' | | |
|---|---|---|---|---|---|---|---|---|---|---|---|
| TP=63 | FP=28 | 91 | TP=77 | FP=77 | 154 | TP=24 | FP=88 | 112 | TP=76 | FP=12 | 88 |
| FN=37 | TN=72 | 109 | FN=23 | TN=23 | 46 | FN=76 | TN=12 | 88 | FN=24 | TN=88 | 112 |
| 100 | 100 | 200 | 100 | 100 | 200 | 100 | 100 | 200 | 100 | 100 | 200 |
| TPR=0.63 | | | TPR=0.77 | | | TPR=0.24 | | | TPR=0.76 | | |
| FPR=0.28 | | | FPR=0.77 | | | FPR=0.88 | | | FPR=0.12 | | |
| ACC=0.68 | | | ACC=0.50 | | | ACC=0.18 | | | ACC=0.82 | | |

Plots of the four results above in the ROC space are given in FIGURE 2-3. The result of **A** clearly shows the best predictive power among **A**, **B** and **C**. The result of **B** lies on the diagonal line, and it can also be seen from TABLE 2-2 that the accuracy of **B** is 50%. However, when **C** is mirrored across the center point (0.5, 0.5), the resulting **C'** is even better than **A**. This mirrored point simply reverses the predictions of whatever method or test produced the **C** contingency table (Confusion matrix), which has positive predictive power. When the **C** 

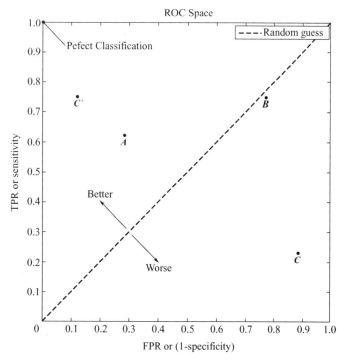

FIGURE 2 – 3  The ROC space

method predicts **P** or **N**, the **C'** method would predict **N** or **P**, respectively. In this manner, the **C'** test would perform better.

## 2.2 Statistical Based Algorithm

### 2.2.1 Bayesian classification

Bayesian classifier is based on applying *Bayes' Theorem* with the "naive" assumption of conditional independence between every pair of features given the value of the class variable. For example, a fruit may be considered to be an apple if it is red, round, and about 10 cm in diameter. A naive Bayes classifier considers each of these features to contribute independently to the probability that this fruit is an apple, regardless of any possible correlations between the color, roundness, and diameter features. This obviously is a very strong assumption.

Bayes' theorem states the following relationship, given a class variable $C_k$ and dependent feature vector $x_1$ through $x_n$:

$$P(C_k | x_1, x_2, \dots, x_n) = \frac{P(C_k) P(x_1, x_2, \dots, x_n | C_k)}{P(x_1, x_2, \dots, x_n)}$$

Note here, in plain expression, the above equation can be written as:

$$Posterior = \frac{Prior \times Likelihood}{Evidence}$$

In practice, there is interest only in the numerator of that fraction, because the denominator does not depend on $C_k$, and the values of the features $x_i$ are given, the denominator is effectively constant.

Using the naive conditional independence assumption that:

$$P(x_i | C_k, x_1, x_2, \dots, x_{i-1}, x_i, x_{i+1}, \dots, x_n) = P(x_i | C_k)$$

for all $i$, this relationship is simplified to:

$$P(C_k | x_1, x_2, \dots, x_n) = \frac{P(C_k) \prod_{i=1}^{n} P(x_i | C_k)}{P(x_1, x_2, \dots, x_n)}$$

Since $P(x_1, x_2, \dots, x_n)$ is constant, and then we get:

$$P(C_k, x_1, x_2, \dots, x_n) \propto P(C_k) \prod_{i=1}^{n} P(x_i | C_k)$$

$$\Downarrow$$

$$\hat{C_k} = \arg \max_{c_k} P(C_k) \prod_{i=1}^{n} P(x_i | C_k)$$

where $\propto$ denotes proportionality.

We can use Maximum A Posteriori (MAP) estimation to estimate $P(C_k)$ and $P(x_i | C_k)$; the former is then the relative frequency of class $C_k$ in the training set. Then, the different naive Bayesian classifiers differ mainly by the assumptions they make regarding the distribution of $P(x_i | C_k)$. This greatly reduces the computational cost by only counting the class distribution. Even though the over-simplified assumption is not valid in most cases since the attributes are dependent, Bayesian classifiers have surprisingly performed quite well in many real-world situations, such as the popular document classification (i.e whether a document belongs to the category of sports, politics, technology, etc. The features/predictors used by the classifier are the frequency of the words present in the document) and spam filtering. They require a small amount of training data to estimate the necessary parameters, which has been proved theoretically. It can be easily scalable to larger datasets since it takes linear time, rather than by expensive iterative approximation as used for many other sophisticated classifiers. On the other hand, the method suffers from the following drawbacks.

- If the categorical variable belongs to a category that wasn't followed up in the training set, then the model will give it a probability of 0 which will stop it from making any prediction (see the EXAMPLE 2-2).
- Naive Bayes assumes independence betweenthe features. In real life, it is difficult to gather data that involves completely independent features.

A simple example is illustrated in the following EXAMPLE 2-2.

 **EXAMPLE 2-2**

Suppose we are building a classifier that says whether a text is about sports or not. Our training data has 5 sentences shown below.

| No. | Text | Tag |
|---|---|---|
| 1 | "A great game" | Sports |
| 2 | "The election was over" | Not sports |
| 3 | "Very clean match" | Sports |
| 4 | "A clean but forgettable game" | Sports |
| 5 | "It was a close election" | Not sports |

Now, which tag does the sentence *A very close game* belong to? Since Naive Bayes is a probabilistic classifier, we want to calculate the probability that the sentence *A very close game* is **Sports** and the probability that it's *Not Sports*. Then, we take the larger one. Here what we expect is $P(Sports \mid a\ very\ close\ game)$——the probability that the label of a sentence is *Sports* given that the sentence is *A very close game*.

So what should we do? We use word frequencies. That is, we ignore word order and sentence construction, treating every document as a set of the words it contains. By using the Bayes theorem, we get:

$$P(Sports \mid a\ very\ close\ game) = \frac{P(a\ very\ close\ game \mid Sports) \times P(Sports)}{P(a\ very\ close\ game)}$$

For calculation, we can discard the denominator $P(a\ very\ close\ game)$ and assume that every word in a sentence is independent of the other ones (the naïve assumption). This means that we do not need to look at entire sentences, but rather at individual words. So we can write $P(a\ very\ close\ game)$ as:

$$P(a) \times P(very) \times P(close) \times P(game).$$

Then $P(a\ very\ close\ game\ |\ Sports)$ can be expressed as:
$$P(a|Sports) \times P(very|Sports) \times P(close|Sports) \times P(game|Sports)$$

Because these individual words (*a*, *very*, *close*, *game*) show up several times in our training data, we can calculate these conditional probabilities as follows:

- First, we get the probability of *Sports* $P(Sports)$ is ⅗. Then, $P(Not\ Sports)$ is ⅖.
- Second, the $P(game\ |\ Sports) = 2/11$ for the word "*game*" appears in Sports texts is 2 times, the total number of words in sports is 11.
- Since the number of possible words is 14, namely, ['a', 'great', 'very', 'over', 'it', 'but', 'game', 'election', 'clean', 'close', 'the', 'was', 'forgettable', 'match'], and applying Laplace smoothing to tackle the problem of word "close" that doesn't appear in any Sports text, we get that $P(close\ |\ Sports) = \frac{0+1}{11+14}$. The full results are as follow.

| Word | $P(word\ |\ Sports)$ | $P(word\ |\ not\ Sports)$ |
|---|---|---|
| A | $\frac{2+1}{11+14}$ | $\frac{1+1}{9+14}$ |
| very | $\frac{1+1}{11+14}$ | $\frac{0+1}{9+14}$ |
| close | $\frac{0+1}{11+14}$ | $\frac{1+1}{9+14}$ |
| game | $\frac{2+1}{11+14}$ | $\frac{0+1}{9+14}$ |

- Finally, we can obtain all the probabilities of the sentence of being **Sports** or **Not Sports**.

$P(a\ |\ Sports) \times P(very\ |\ Sports) \times P(close\ |\ Sports) \times$
$P(game\ |\ Sports) \times P(Sports) = 2.76 \times 10^{-5}$
$P(a\ |\ Not\ Sports) \times P(very\ |\ Not\ Sports) \times P(close\ |\ Not\ Sports) \times$
$P(game\ |\ Not\ Sports) \times P(Not\ Sports) = 2.76 \times 10^{-5}$

- So the value of $P(Sports\ |\ a\ very\ close\ game)$ is larger than that of $P(Not+sports\ |\ a\ very\ close\ game)$, the Bayesian classifier gives "*A very close game*" the **Sports** label.

### 2.2.2 Logistic classification

*Logistic* classification (or logit model) is a machine learning algorithm which is used to assign observations to a discrete set of classes. Note that logistic classification is often called *logistic regression* model. However, we prefer to use the term classification here because in a logit model the output is discrete other than continuous. Unlike linear regression which outputs continuous values, *logistic* classification transforms outputs using *logistic* function to return a probability value which can then be mapped to two or more discrete classes. For instance, if we want to speculate about the relationship between the time spent studying and final exam scores, linear regression, and *logistic* regression can make different predictions.

- Linear regression could predict the student's exam score on a scale of $0 \sim 100$. The predictions are continuous (numbers in a range).
- *Logistic classification* is used to predict whether the student passed or failed by viewing the probability scores underlying the model's classifications.

Based on the number of categories, *Logistic* classification can be classified as:

- Binomial: target variable has only 2 possible types: "0" or "1" which may represent "win" vs "loss" "pass" vs "fail" "yes" vs "no", etc.
- Multinomial: target variable has 3 or more possible types which are not ordered (i.e. types have no quantitative significance) like "dog" vs "cat" vs "fish".
- Ordinal: it handles target variables with ordered categories. For example, an exam score can be categorized as: "very poor" "poor" "good" "very good". And each class can be given a score like 0, 1, 2, 3.

When it comes to classification, one should determine the probability of a sample being part of a predefined class or not. Therefore, it is expected that the probability should be in a range of 0 and 1. In practice, the **sigmoid function** is used to map any real predicted values to probabilities between 0 and 1, which has s common S-shape curve with the equation $\sigma(x) = \frac{1}{1+e^{-x}}$. Curve of a standard sigmoid function shown in FIGURE 2-4.

We can infer from the above figure that:
- $\sigma(x)$ tends towards 1 as $x \to \infty$.
- $\sigma(x)$ tends towards 0 as $x \to -\infty$.
- $\sigma(x)$ is always bounded between 0 and 1.

Suppose given a sample $(y_i, x_i)$ for $i = 1, 2, \ldots, N$, each observation in the sample contains:

- Target variable denoted by $y_i$. In the case of a binary logit model, $J = 2$, $y_1 = 1$ and $y_1 = 0$ (it is a Bernoulli random variable). For a multinomial logit model, $J \geqslant 2$.
- An input of $1 \times K$ vector, denoted by $x_i$.

The logit model can be viewed as a latent variable model, i.e., an unobserved variable

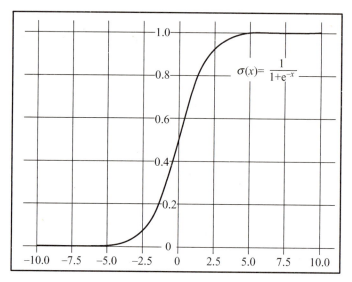

FIGURE 2-4 Curve of a standard sigmoid function

$z_i = \mathbf{x}_i^T \beta + \xi_i$, where $\xi_i$ is a random error term that adds noise to the relationship between the inputs $\mathbf{x}_i$ and $z_i$, and $\beta = (\beta_1, \beta_2, \ldots, \beta_K)^T$ is the regression coefficient vector. The latent variable $z_i$ then can be used to determine $y_i$:

$$y_i = \begin{cases} 1 & \text{if } z_i \geq 0 \\ 0 & \text{if } z_i < 0 \end{cases}$$

Then the conditional probabilities for binary labels (0 and 1) for the $i^{th}$ sample is written as:

$$\begin{aligned} P(y_i=1 \mid x_i; \beta) &= P(z_i \geq 0 \mid x_i) \\ &= P(\mathbf{x}_i^T \beta + \xi_i \geq 0 \mid x_i) \\ &= P(\xi_i \geq -\mathbf{x}_i^T \beta \mid x_i) \\ &= P(\xi_i \leq \mathbf{x}_i^T \beta \mid x_i) \text{ (for the symmetric distribution)} \\ &= F(\mathbf{x}_i^T \beta) \end{aligned}$$

$F(\ )$ is the cumulative distribution function of the error $\xi_i$ which has a standard sigmoid distribution. If we use different distributions of $\xi_i$, we can get some other classification models, such as *probit* model when $\xi_i$ has a standard normal distribution.

Now it is the time to determine the decision boundary. In practice, to map a probability score to a discrete class, we usually set a threshold value or tipping point, such as 0.5, above which we will classify samples into class 1 and below which we classify values into class 0. For logistic classification with multiple classes, we could select the class with the highest predicted probability. As the probability gets closer to 1, we are more confident to say that the observation is in class 1. We illustrate this in EXAMPLE 2-3 for detail.

## EXAMPLE 2-3

A group of 20 students spend 0-6 hours each day studying to pass an exam, the sample data is in the below table. What is the effect of the number of studied hours spent on the probability of passing the exam?

| Studied_hours | 0.50 | 0.75 | 1.00 | 1.25 | 1.50 | 1.75 | 1.75 | 2.00 | 2.25 | 2.50 |
|---|---|---|---|---|---|---|---|---|---|---|
| Pass | 0(failed) | 0 | 0 | 0 | 0 | 0 | 1 | 0 | 1 | 0 |
| Studied_hours | 2.75 | 3.00 | 3.25 | 3.50 | 4.00 | 4.25 | 4.50 | 4.75 | 5.00 | 5.50 |
| Pass | 1(passed) | 0 | 1 | 0 | 1 | 1 | 1 | 1 | 1 | 1 |

First, let us build the linear regression equation of the latent variable with studied hours, i.e., $z = \beta_0 + \beta_1 Studied\_hours$, then we transform the target variable using the sigmoid function into a probability value ranging from 0 to 1:

$$P(Pass=1) = \frac{1}{1+e^{-z}}$$

$$= \frac{1}{1+e^{-\beta_0 + \beta_1 Studied\_hours}}$$

By using an efficient estimation approach (such as maximum likelihood estimation using cross-entropy loss functions), the coefficients $\beta_0$ and $\beta_1$ can be found as $-4.078$ and $1.505$, respectively. So for a student who spends 2 hours on study, the estimated probability of passing the exam is 0.256, which indicates there is a 25.6% chance of passing the exam. If the decision boundary was set at 0.5, we would classify this observation as "Fail". Similarly, if a student spends 4 hours on study, he/she can have an 87.5% chance of passing the exam, then we classify this observation as "Pass".

## 2.3 Decision Tree-based Algorithm

### 2.3.1 What is a decision tree

The decision tree is one of the most popular tools in the family of classification, which makes use of a tree-like structure to deliver consequences based on the input dataset (the training data). Each internal (non-leaf) node denotes a test on an attribute, each branch represents the outcome of a test, and each leaf (or terminal) node holds a class label. The topmost node in a tree is the root node. Decision tree is capable of handling heterogeneous as well as missing data, and is further capable of producing understandable rules that follow an "*if Variable A is X then...*" pattern. One of the easiest models to interpret is focused on linearly separable data. If you can't draw a straight line through it, basic implementations of decision trees aren't as useful, more complicated algorithms would be required.

A definition for a decision tree could be stated as: given a database $D = \{t_1, t_2, \ldots, t_n\}$ of tuples (items, records), where $t_i = (t_{i1}, t_{i2}, \ldots, t_{im})$ contains the attributes $a = (a_1, a_2, \ldots, a_m)$. A set of classes $C = \{c_1, c_2, \ldots, c_k\}$. A decision tree (DT) is a tree associated with $D$ that has the following properties:

- Each internal node is labeled with an attribute $a_i$.
- Each arc is labeled with a *predicate* that can be applied to the attribute associated with the patent.
- Each leaf node is labeled with a class $c_k$.

Therefore, DT can have three main parts:

- Root Node: The node that performs the first split.
- Terminal Nodes/Leaf Nodes: Nodes that predict the outcome.
- Arcs/Branches: Arrows connecting nodes shows the flow from question to answer.

Two DT examples are visualized as follows shown in FIGURE 2-5. The root node is the starting point of the tree, and both root and terminal nodes contain questions or criteria to be answered. Each node typically has two or more nodes extending from it. Attributes in the database schema that will be used to label nodes which the divisions will take place in the tree and around are called *splitting attributes*. The predicates by which the arcs in the tree are marked are called *splitting predicates*. An illustration structure of decision trees shown in FIGURE 2-5, the *splitting attributes* are {Am I out of shampoo? Is it raining} and {Gender, Height}. The *splitting predicates* for {Am I out of shampoo?} are {= yes, = no}. The *splitting predicates* for {Gender} are {= Female, = Male}.

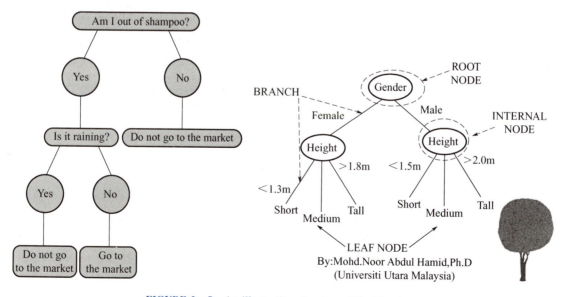

FIGURE 2-5  An illustration structure of decision trees

Here are some pros and cons of the decision tree.
- The pros:
  (1) They are interpretable. People can understand decisions.
  (2) Easily handles irrelevant attributes (information gain=0).
  (3) Requires little data preprocessing: no need for one-hot encoding, dummy variables, and so on.
- The cons:
  (1) They are unstable, meaning that a small change in the data can lead to a large change in the structure of the optimal decision tree.
  (2) They are likely to overfit noisy data. The probability of overfitting on noise increases as a tree gets deeper.
  (3) For data including categorical variables with a different number of levels, information gain in decision trees is biased in favor of those attributes with more levels.

### 2.3.2 Decision tree algorithm

Decision tree algorithm fall into the category of supervised learning algorithms (having a predefined target variable). This contrasts with clustering and similar unsupervised learning tasks where the categories are derived from training data instead of a prior definition. It works for both continuous as well as categorical target variables. Notable ones include:
- ID3 (Iterative Dichotomiser 3). It was developed in 1986 by Ross Quinlan. The algorithm creates a multiway tree, finding for each node (i.e. in a greedy manner) the categorical feature that will yield the largest information gain for categorical targets. Trees are grown to their maximum size and then a pruning step is usually applied to improve the ability of the tree to generalize to unseen data.
- C4.5 and C5.0. It is the successor to ID3 and removed the restriction that features must be categorical by dynamically defining a discrete attribute (based on numerical variables) that partitions the continuous attribute value into a discrete set of intervals. C4.5 converts the trained trees (i.e. the output of the ID3 algorithm) into sets of if-then rules. The accuracy of each rule is then evaluated to determine the order in which it should be applied. Pruning is done by removing a rule's precondition if the accuracy of the rule improves without it. C5.0 is Quinlan's latest version released under a proprietary license. It uses less memory and builds smaller rulesets than C4.5 while being more accurate.
- CART. It is very similar to C4.5, but it differs in that it supports numerical target variables (regression) and does not compute rule sets. CART constructs binary trees using the feature and threshold that yield the largest information gain at each node.

- Chi-square Automatic Interaction Detection (CHAID). Performs multi-level splits when computing classification trees.
- MARS: extends decision trees to handle numerical data better.
- Conditional inference trees. A statistics-based approach that uses non-parametric tests as splitting criteria, corrected for multiple testing to avoid overfitting. This approach results in unbiased predictor selection and does not require pruning.
- Other algorithms, like random forest and gradient boosting, are also being popularly applied in all kinds of data mining problems.

Note that the challenge facing any inductive learning algorithm is to produce a tree that both covers all the training examples correctly, and has the highest probability of being correct on new instances. A valuable heuristic for inducing decision trees comes from the time-honored logical principle of Occam's Razor which was first articulated by the medieval logician, William of Occam, holds that one should always prefer the simplest correct solution to any problem. In the above case, this would favor decision trees that not only classify all training examples but also examine properties as fewer as possible. The reason is straightforward: the simplest decision tree that correctly handles the known examples is the tree that makes the fewest assumptions about unknown instances. In other words, the fewer assumptions made, the less likely we are to make mistakes.

These above mentioned recursive algorithms build a tree in a top-down pattern by checking training data. Using the initial training data, the "best" splitting attribute is chosen first. Algorithms differ in how they determine the "best attribute" and its "best predicates" to use for splitting. Once this has been determined, the node and its arcs are created and added to the created tree. The algorithm continues recursively by adding new subtrees to each branching arc, and terminates when some "stopping criteria" is reached. Again, the criteria of each algorithm to determine when to stop the tree differs. One simple approach for stopping is when the tuples in the reduced training set all belong to the same class. This class is then used to label the leaf node created. In summary, a decision tree usually follows these steps.

- **Root node:** Scan each variable and try to split the data based on the important features by using different metrics such as Entropy, Information Gain, and Gini index, etc.
- **Split:** Given some splitting criterion, compare each split and see which one performs best.
- **Repeat:** The two steps above for each node, considering only the data points in each node.
- **Pre-Pruning:** Stop partitioning if the node is too small to be split, any future partition is too small, or cross-validation data shows worse performance.
- **Post-Pruning:** Alternatively, trim the decision tree by removing duplicate rules or simplifying.

### 2.3.3 Metrics of selecting splitting points

Algorithms for constructing decision trees usually work top-down, by choosing an attribute at each step that best splits the set of items. There are several different criteria for measuring the "best". In this text, we present three popular metrics, namely, Gini Index, Information Gain, and Information Gain Ratio.

(1) **Gini Index,** or Gini impurity, favors large partitions, calculated by subtracting the sum of the squared probabilities of each class from one. It is a measure of how often a randomly chosen element from the set would be incorrectly labeled if it was randomly labeled according to the distribution of labels in the subset, used by the CART algorithm for classification trees.

Let $p_i$ is the frequency of class $c_i$, then the mathematical formulation of the Gini Index is:

$$\text{Gini} = 1 - \sum_{i=1}^{c} p_i^2$$

Then the goodness of a split of a database ($D$) with $n$ observations into subsets $D_1$ and $D_2$, with $n_1$ and $n_2$ observations, respectively, is defined by

$$\text{Gini}(D) = \frac{n_1}{n}\text{Gini}(D_1) + \frac{n_2}{n}\text{Gini}(D_2)$$

So we choose the split with the best Gini value.

(2) **Information Gain** is based on the concept of Shannon entropy and information content from information theory. It favors smaller partitions with many distinct values, and is used by the ID3 tree-generation algorithms. Entropy is a measure of randomness in information processing, and the higher the entropy, the more difficult it is to draw any conclusions from the information.

Entropy is defined as below

$$H(D) = H(p_1, p_2, \dots, p_k) = -\sum_{i=1}^{K} p_i \log_2 p_i$$

where $p_1, p_2, \dots, p_K$ are fractions that add up to 1 and represent the percentage of each class in the child node that results from a split in the tree. Note $\log_2^0 = 0$ in entropy calculations.

Given a training dataset, $D$, the entropy $H(D)$ finds the randomness of $D$. When $D$ is split into $S$ new subsets $S = \{D_1, D_2, \dots, D_s\}$ by the $a^{th}$ attribute $x_a$ taking a value $x_a^u$. $s \in S$, we can again check the entropy of those subsets $H(D_i) = H(D \mid x_a^u(i), i \in S)$. The Information Gain can be calculated in terms of Shannon entropy:

$$IG(D, S) = H(D) - \sum_{i \in S} \frac{|D_i|}{|D|} H(D \mid x_a^u(i), i \in S)$$

Consider an example dataset with four attributes: outlook (sunny, overcast, rainy), temperature (hot, mild, cool), humidity (high, normal), and windy (true, false), with a binary (yes or no) target variable, *play*, and 14 data points. To construct a decision tree on this data, we need to compare the information gain of each of four trees, each split on one of the four features. The split with the highest Information Gain will be taken as the first split and the process will continue until all children nodes are pure, or until the Information Gain is zero.

The split using the feature windy results in two children nodes, 1 for a windy value of true and 1 for a windy value of false. In this dataset, there are 6 data points with a true windy value, 3 of which have a play (where play is the target variable) value of yes and 3 with a play value of no. The eight remaining data points with a windy value of false contain 2 "no's" and 6 "yes's".

The information of the windy = true node is calculated using the entropy equation above. Since there is an equal number of yes's and no's in this node, we have:

$$H([3,3]) = -\frac{3}{6}\log_2\frac{3}{6} - \frac{3}{6}\log_2\frac{3}{6} = -\frac{1}{2}\log_2\frac{1}{2} - \frac{1}{2}\log_2\frac{1}{2} = 1$$

For the node where windy=false there were eight data points, 6 yes's and 2 no's. Thus we have:

$$H([6,2]) = -\frac{6}{8}\log_2\frac{6}{8} - \frac{2}{8}\log_2\frac{2}{8} = -\frac{3}{4}\log_2\frac{3}{4} - \frac{1}{4}\log_2\frac{1}{4} = 0.81$$

To find the information of the split, we take the weighted average of these two numbers based on how many observations fell into which node.

$$H([3,3],[6,2]) = H(\text{windy or not}) = 0.89$$

To find the Information Gain of the split using windy, we must first calculate the information in the data before the split. The original data contained 9 yes's and 5 no's.

$$H([9,5]) = -\frac{9}{14}\log_2\frac{9}{14} - \frac{5}{14}\log_2\frac{5}{14} = 0.94$$

Based on the above, we now can calculate the information gain achieved by splitting on the windy feature.

$$IG(\text{windy}) = H([9,5]) - H([3,3],[6,2]) = 0.94 - 0.89 = 0.05$$

To build the tree, the information gain of each possible first split would need to be calculated. The best first split is the one tha thas the most Information Gain. This process is repeated for each impure node until the tree is complete.

Although information gain is usually a good measure for deciding theimportance of an attribute, it is not perfect. A notable problem occurs when information gain is applied to attributes that can take on a large number of distinct values. For example, suppose that one is building a decision tree for some data describing the customers of a business. Information gain is often used to decide which of the attributes are the most relevant, so they can be tested near the root of the tree. One of the input attributes might be the customer's credit card number. This attribute has high mutual information, because it uniquely identifies each customer, but we do not want to include it in the decision tree; deciding how to treat a customer based on their credit card number is unlikely to generalize to customers we haven't seen before (which is overfitting).

(3) **Information Gain Ratio** is a ratio of information gain to the intrinsic information. It was proposed by Ross Quinlan, to reduce bias towards multi-valued attributes by taking the number and size of branches into account when choosing an attribute. When ID3 is applied to some cases with attributes of many segmentation, it may lead to overfitting issues. For an extreme instance, an attribute that has a unique value for each tuple in the training dataset would be the best because there would be only one tuple (and thus one class) for each split. An improvement can be made by using the **Information Gain Ratio** rather than **Information Gain**. The definition is

$$\text{GainRatio} = IG(D,S)/H\left(\frac{|D_1|}{|D|},\frac{|D_2|}{|D|},\cdots,\frac{|D_n|}{|D|}\right)$$

C4.5 uses the largest Information Gain Ratio which can ensure a larger than average Information Gain. Information Gain Ratio biases the decision tree against considering attributes with a large number of distinct values. So it solves the drawback of Information Gain. For example, suppose that we are building a decision tree for some data describing a business's customers. Information Gain is often used to decide which of the attributes are the most relevant, so they can be tested near the root of the tree. One of the input attributes might be the customer's credit card number. This attribute has a high Information Gain because it uniquely identifies each customer, but we do not want to include it in the decision tree; deciding how to treat a customer based on their credit card number is unlikely to generalize to customers we haven't seen before (i.e. overfitting). It is worth pointing out that, in practice, you should experiment with your own dataset and the splitting criteria.

### 2.3.4 ID3 and C4.5

ID3 begins with the original dataset $D$ as the root node. On each iteration of the algorithm, it iterates through the very unused attribute of the dataset $D$ and calculates Entropy ($H$) and Information Gain($IG$) of the attribute. It then selects the attribute which has the smallest Entropy or Largest Information Gain. The dataset $D$ is then split by the selected attribute to produce a subset of the dataset. The algorithm continues to recur on each subset, considering only attributes never selected before. Throughout the algorithm, the decision tree is constructed with each *non-Terminal* node (internal node) representing the selected attribute on which the data was split, and *Terminal* nodes (Leaf Nodes) representing the class label of the final subset of this branch.

C4.5 builds decision trees from a set of training data in the same way as ID3 by using the concept of information entropy. The attribute with the highest normalized Information Gain is chosen to make the decision. C4.5 has a number of improvements over ID3 in several ways, such as:

- It can handle both continuous and discrete attributes. To handle continuous attributes, C4.5 creates a threshold and then splits the dataset into two parts. In each part, the attribute value is either above the threshold or less than or equal to the threshold.
- It can handle training data with missing attribute values. Missing attribute values are simply ignored in gain and entropy calculations.
- Pruning trees after creation-C4.5 goes back through the tree once it's been created and attempts to remove branches that do not help by replacing them with leaf nodes.

### 2.3.5 Classification and regression trees (CART)

CART is a non-parametric decision tree learning algorithm that produces either classification or regression trees, depending on whether the dependent variable is categorical or numeric, respectively. It was introduced in 1984 by Leo Breiman, Jerome Friedman, Richard Olshen and Charles Stone as an umbrella term to refer to the following types of decision trees:

- **Classification Trees**: where the target variable is categorical and the tree is used to identify the "class" within which a target variable would likely fall into.
- **Regression Trees**: where the target variable is continuous and the tree is used to predict its value.

Some useful features and advantages of CART:

- Nonparametric and does not rely on datawith a particular type of distribution.
- Not significantly impacted by outliers in the input variables.
- Stopping rules can be relaxed to "overgrow" decision trees and then prune back the tree to the optimal size, which minimizes the probability that important structures in the dataset will be overlooked by stopping too soon.
- Use the same variables more than once in different parts of the tree. This capability can uncover complex interdependencies between sets of variables.

### 2.3.6 Model overfitting

Overfitting happens when a model memorizes its training data so well, and may, therefore, fail to fit additional data or predict future observations reliably. It is one of the nontrivial challenges faced while using decision-tree based algorithms. If there is no limit set or constraints of a decision tree, it will give you 100% accuracy on the training set because in the worse case it will end up making one leaf for each sample. Thus, preventing overfitting is pivotal while modeling a decision tree. Generally, there are two ways we can count on, i. e. , *setting constraints on tree size*, and *tree pruning*. We discuss them briefly in the following sections.

(1) Setting constraints on tree-based algorithms

This can be done by using various parameters that are used to define a tree. First, let's look at the general structure of a decision tree again in FIGURE 2-6. The parameters described below are based on the structure of a decision tree, irrespective of the tool. It is important to understand the role of parameters used in tree modeling. These parameters are available in the corresponding packages in **R & Python**.

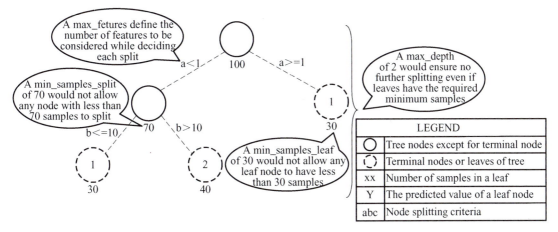

FIGURE 2-6　The general structure of a decision tree

- **Minimum samples for a node split**
    - √ Defines the minimum number of samples (or observations) which are required in a node to be considered for splitting.
    - √ Used to control over-fitting. Higher values prevent a model from learning relations which might be highly specific to the particular sample selected for a tree.
    - √ Too high values can lead to under-fitting hence, it should be tuned using cross-validation.
- **Minimum samples for a terminal node (leaf node)**
    - √ Defines the minimum samples (or observations) required in a terminal node or leaf nodes.
    - √ Used to control over-fitting similar to *min_samples_split*.
    - √ Generally, lower values should be chosen for imbalanced class problems because the regions in which the minority class will be in majority will be very small.

- **Maximum depth of the tree (vertical depth)**
    - √ The maximum depth of a tree.
    - √ Used to control over-fitting as higher depth will allow the model to learn relations very specific to a particular sample.
    - √ Should be tuned using cross-validation.
- **Maximum number of terminal nodes**
    - √ The maximum number of terminal nodes or leaf nodes in a tree.
    - √ It can be defined in place of *max_depth*. Since binary trees are created, a depth of '$n$' would produce a maximum of $2^n$ leaves.
- **Maximum features to consider for a split**
    - √ The number of features to consider while searching for the best split. These will be randomly selected.
    - √ As a thumb-rule, square root of the total number of features works great but we should check up to 30%~40% of the total number of features.
    - √ Higher values can lead to over-fitting but depend on case to case.

(2) Pruning

Pruning is a technique that reduces the size of decision trees by removing sections of the tree that provide little power to classify observations. Pruning reduces the complexity of the final classifier, and hence improves predictive accuracy by the reduction of overfitting.

One of the simplest forms of pruning is reduced error pruning. Starting at the leaves, each node is replaced with its most popular class. If the prediction accuracy is not affected then the change is kept. While somewhat naive, reduced error pruning has the advantage of simplicity and speed.

Minimal cost-complexity pruning provides another option to control the size of a tree, which is parameterized by $\alpha \geqslant 0$ called the complexity parameter. It is used to define the cost-complexity measure of a given tree $T$, $C_\alpha(T) = C(T) + \alpha |T|$, where $|T|$ is the number of Terminal Nodes in $T$ and $C(T)$ is usually defined as the total misclassification rate (error rate) of the Terminal Nodes. Alternatively, $C(T)$ can be replaced with the total sample weighted impurity of the terminal nodes, as used in *scikit-learn* package in Python. Greater values of $\alpha$ increases the number of nodes pruned. Minimal cost-complexity pruning finds the subtree of $T$ that minimizes $C_\alpha(T)$. The cost complexity measure of a single node is $C_\alpha(t) = C(t) + \alpha$. The branch, $T_t$, is defined to be a tree where node $t$ is its root. In general, the impurity of a node is greater than the sum of impurities of its terminal nodes, $C(T_t) < C(t)$. However, the cost complexity measure of a node, $t$, and its branch, $T_t$, can be equal depending on $\alpha$. We define the effective $\alpha$ of a node to be the value where they are equal, $C_\alpha(T_t) = C_\alpha(t)$ or $\alpha_{eff}(t) = \dfrac{C(t) - C(T_t)}{|T| - 1}$. A non-terminal node with the smallest value of $\alpha_{eff}(t)$ is the weakest link and will be pruned.

### 2.3.7 Visualize a decision tree in Python

Here we give the Python code of visualizing a decision tree.

(1) Preliminaries

```python
# Load libraries
from sklearn.tree import DecisionTreeClassifier
from sklearn import datasets
from IPython.display import Image
from sklearn import tree
import pydotplus
```

(2) Load iris data

```python
# Load data
iris = datasets.load_iris()
X = iris.data
y = iris.target
```

(3) Train decision tree

```python
# Create decision tree classifer object
clf_tree = DecisionTreeClassifier(random_state=123)
# Train model
model = clf_tree.fit(X, y)
```

(4) Visualize decision tree

```python
# Create DOT data
dot_data = tree.export_graphviz(clf_tree, out_file=None,
                                feature_names=iris.feature_names,
                                class_names=iris.target_names)
# Draw graph
graph = pydotplus.graph_from_dot_data(dot_data)
# Decision tree Visualization with the Iris dataset, as in FIGURE 2-7
Image(graph.create_png())
# Create PDF
graph.write_pdf("iris.pdf")
# Create PNG
graph.write_png("iris.png")
#
```

(5) Save the decision tree image to a image file

```python
# Create PDF
graph.write_pdf("iris.pdf")
# Create PNG
graph.write_png("iris.png")
```

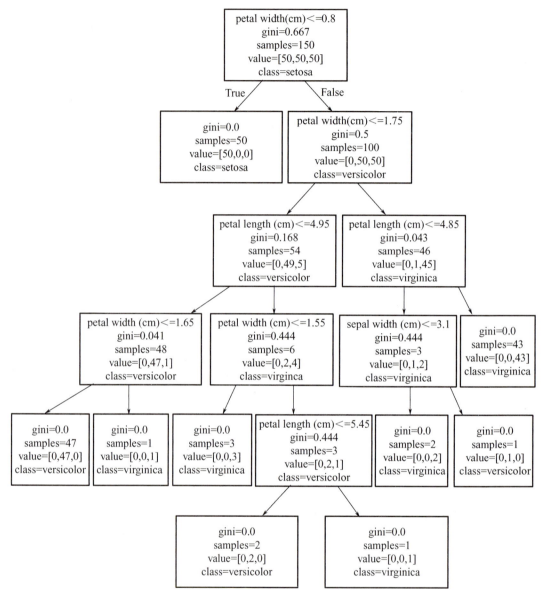

FIGURE 2-7  Decision tree visualization with the Iris dataset

## 2.4  Rule-based Classification

Rule-based classifier makes use of a set of IF-THEN rules for classification. We can express a rule in the following form:

IF *condition* THEN *conclusion*

Let us consider a rule R1:

IF *age*=*youth* AND *student*=*yes*

THEN *buy_computer*=*yes*

The IF part of the rule is called rule *antecedent or precondition*. The THEN part of the rule is called *rule consequent*. The antecedent part of the condition consists of one or more attribute tests and these tests are logically ANDed. The consequent part consists of class prediction. We can also write rule R1 as follows:

$$(age = youth) \wedge (student = yes)) \rightarrow (buys\ computer = yes)$$

If the condition holds true for a given tuple, then the antecedent is satisfied. Coverage of a rule is defined as the fraction of records that satisfy the antecedent of a rule. Accuracy of a rule is defined as the fraction of records that satisfy both the antecedent and consequent of a rule.

### 2.4.1 Generating rules from a DT

The process to generate a rule from a DT is straightforward. All rules with the same consequent could be combined by ORing the antecedents of the simpler rules.

Although a DT can be used to generate rules, they are not the same. Some differences between rules and trees include:
- The tree has an implied order in which the splitting is performed. Rules have no order.
- A tree is created based on checking all classes. Only one class must be examined at a time when generating rules.

If we convert the result of the decision tree to classification rules, these rules would be mutually exclusive and exhaustive at the same time. For mutually exclusive rules, classifier contains mutually exclusive rules if the rules are independent of each other, and every record is covered by at most one rule. For exhaustive rules, the classifier has exhaustive coverage if it accounts for every possible combination of attribute values, and each record is covered by at least one rule. These rules can be simplified. However, simplified rules may no longer be mutually exclusive or exhaustive since a record may trigger more than one rule.

### 2.4.2 One rule (1R) algorithm

1R is one of the simple approaches. It generates a simple set of rules that are equivalent to a DT with only one level. Choosing the best attribute to perform the classification based on the training data is the basic idea of 1R. "Best" is defined by counting the number of errors. 1R is illustrated in TABLE 2-3. Suppose we have a decision tree based on attributes of *Gender* and *Height*. If we only use the *Gender* attribute, there are a total of 6/15 errors, whereas if we use the *Height* attribute, there are only 1/15. In this case, the *Height* would be chosen and the six rules stated in the table would be used.

**TABLE 2-3  1R illustrate**

| Option | Attribute | Rules | Errors | Total Errors |
|---|---|---|---|---|
| 1 | Gender | F → Medium | 3/9 | 6/15 |
|   |        | M → Tall | 3/6 |  |
| 2 | Height | (0, 1.6] → Short | 0/2 | 1/15 |
|   |        | (1.6, 1.7] → Short | 0/2 |  |
|   |        | (1.7, 1.8] → Medium | 0/3 |  |
|   |        | (1.8, 1.9] → Medium | 0/4 |  |
|   |        | (1.9, 2.0] → Medium | 1/2 |  |
|   |        | (2.0, ∞) → Tall | 0/2 |  |

As ID3, 1R tends to choose attributes with a large number of values leading to overfitting. By adding an additional attribute value for the value of *missing*, 1R can handle missing data.

## 2.5  Some Advanced Classification Algorithm

### 2.5.1  Support vector machine algorithm

Support Vector Machine (SVM) belong to a family of generalized linear classifiers and can be interpreted as an extension of the perceptron. They can also be considered a special case of Tikhonov regularization. A special property is that they simultaneously minimize the empirical classification error and maximize the geometric margin; hence they are also known as maximum margin classifiers. Samples on the margin are called the support vectors. SVM algorithm see FIGURE 2-8. A comparison of the SVM with other classifiers has been made by Meyer, Leisch and Hornik.

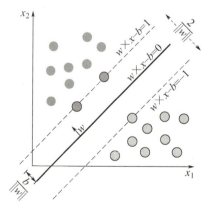

**FIGURE 2-8  SVM algorithm**

Potential drawbacks of the SVM include the following aspects:
- Requires full labeling of input data.
- Uncalibrated class membership probabilities——SVM stems from Vapnik's theory which avoids estimating probabilities on finite data.
- The SVM is only directly applicable for two-class tasks. Therefore, algorithms that reduce the multi-class task to several binary problems have to be applied.
- Parameters of a solved model are difficult to interpret.

### 2.5.2 Random forest algorithm

Random forest classifiers are a type of ensemble learning method that is used for classification, regression and other tasks that can be performed with the help of the decision trees. These decision trees can be constructed at the training time and the output of the class can be either classification or regression. With the help of these random forests, one can correct the habit of overfitting to the training set. Some of the advantages and disadvantages of random forest classifiers are as follows.
- Advantages: random forest classifiers facilitate the reduction in the overfitting of the model and these classifiers are more accurate than the decision trees in several cases.
- Disadvantages: random forest exhibits real-time prediction but that is slow in nature. They are also difficult to implement and have a complex algorithm.

# EXERCISES

1. Using the data for height classification in the following table, construct a confusion matrix assuming output2 is the correct assignment, and output1 is what is actually made.

| Name | Gender | Height | Output1 | Output2 |
|---|---|---|---|---|
| Kristina | F | 1.6m | Short | Medium |
| Jim | M | 2m | Tall | Medium |
| Maggie | F | 1.9m | Medium | Tall |
| Martha | F | 1.88m | Medium | Tall |
| Stephanie | F | 1.7m | Short | Medium |
| Bob | M | 1.85m | Medium | Medium |
| Kathy | F | 1.6m | Short | Medium |
| Dave | M | 1.7m | Short | Medium |
| Worth | M | 2.2m | Tall | Tall |
| Steven | M | 2.1m | Tall | Tall |
| Debbie | F | 1.8m | Medium | Medium |

continued

| Name | Gender | Height | Output1 | Output2 |
|---|---|---|---|---|
| Todd | M | 1.95m | Medium | Medium |
| Kim | F | 1.9m | Medium | Tall |
| Amy | F | 1.8m | Medium | Medium |
| Wynette | F | 1.75m | Medium | Medium |

2. Using 1R, generate rules for exercise 1 of the height classification using the output2 column as the class.

3. Calculate information Gain and Gini index by hand, supposing we have a dataset shown below.

| No. | Var1 | Var2 | Class |
|---|---|---|---|
| 1 | 0 | 33 | A |
| 2 | 0 | 54 | A |
| 3 | 0 | 56 | A |
| 4 | 0 | 42 | A |
| 5 | 1 | 50 | A |
| 6 | 1 | 55 | B |
| 7 | 1 | 31 | B |
| 8 | 0 | −4 | B |
| 9 | 1 | 77 | B |
| 10 | 0 | 49 | B |

4. Assume we wish to assign a credit risk of high, moderate, or low to people based on the following properties of their credit rating:

- Collateral, with possible values {adequate, none}.
- Income, with possible values {"$0 to $15K", "$15K to $35K", "over $35K"}.
- Debt, with possible values {high, low}.
- Credit History, with possible values {good, bad, unknown}.

Try to generate a decision tree using the ID3 algorithm with the table below.

| No. | Credit History | Debt | Collateral | Income | Income_code | Risk |
|---|---|---|---|---|---|---|
| 1 | bad | high | none | $0-15K | 0 | high |
| 2 | unknown | high | none | $15-35K | 1 | high |
| 3 | unknown | low | none | $15-35K | 1 | moderate |
| 4 | unknown | low | none | $0-15K | 0 | high |
| 5 | unknown | low | none | over $35K | 2 | low |
| 6 | unknown | low | adequate | over $35K | 2 | low |

continued

| No. | Credit History | Debt | Collateral | Income | Income_code | Risk |
|---|---|---|---|---|---|---|
| 7 | bad | low | none | $0 to $15K | 0 | high |
| 8 | bad | low | adequate | over $35K | 2 | moderate |
| 9 | good | low | none | over $35K | 2 | low |
| 10 | good | high | adequate | over $35K | 2 | low |
| 11 | good | high | none | $0 to $15K | 0 | high |
| 12 | good | high | none | $15 to $35K | 1 | mod |
| 13 | good | high | none | over $35K | 2 | low |
| 14 | bad | high | none | $15 to $35K | 1 | high |

本章配套资源

# Chapter 3

# Clustering

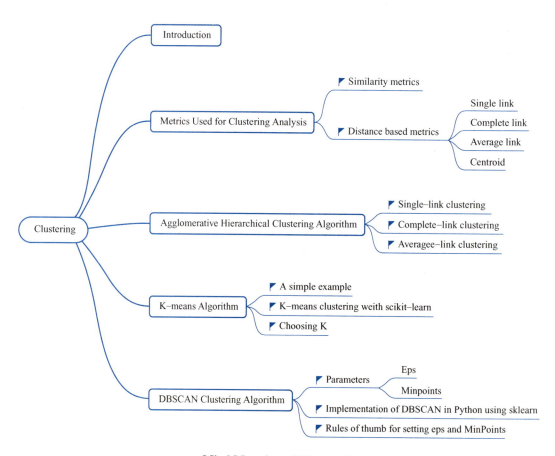

**Mind Mapping of Chapter 3**

聚类分析类似于分类技术，但这里的类或组不是预定义的。这些类或组被称为聚类，或被视为特殊类型的分类或分割。它是一种无监督学习方法，其任务是将总体或数据集划分为多个组，以便同一组中的数据点比其他组中的数据点更相似，是基于它们之间的相似性和不相似性的对象的集合。客户细分的简单示例可以在示例 3-1 中找到。假设您正在管理一家租赁商店，并希望了解客户的偏好以扩展您的业务，您是否可以通过查看每个客户的购买情况来为每个客户制定特定的促销策略？当客户数量非常大时，这绝对不是一种明智之举。事实上，您可以根据他们的购买习惯和个人属性将所有客户分为 10 个组，然后对这 10 个组中的客户采取单独的策略。

聚类分析是一种通用工具，是用于深入了解数据分布以观察每个聚类的特征。聚类分析可以在许多领域提供帮助，比如生物学等。聚类分析可用于识别具有类似行为的某个记录的客户，聚类分析还可以帮助广告商在其客户群中找到不同的群体。在生物学中，它用于确定植物和动物分类学，用于对具有相似功能的基因进行分类，以及用于洞察种群的固有结构。在地理观测数据库中，聚类分析还可以更容易地找到土地上具有类似用途的区域，如有助于按房屋的类型、价值和目的地识别房屋和公寓组；Web 上的文档聚类也有助于发现隐藏信息。此外，聚类分析还有许多其他应用，包括推荐引擎、社交网络分析、搜索结果分组和异常检测等。与分类技术相比，聚类分析的基本特征包括：①聚类的数量是未知的；②与聚类相关的先验知识不多，如我们只有一组未标记的输入数据，但不知道输出会是什么；③聚类结果是动态的，即用户可以随着偏好的变化从一个聚类切换到另一个聚类。聚类问题定义可以表述为：给定一个元组数据库 $D=\{t_1, t_2,…, t_n\}$，其中 $t_i=\{t_{i1}, t_{i2},…, t_{ip}\}$，$p$ 是属性的维度，$k$ 要生成的聚类数，则聚类问题是定义一个映射函数 $f: D \rightarrow \{1, 2,…,k\}$，其中每个是 $t_i$ 都分配给一个类 $K_j$，$1 \leqslant j \leqslant k$，也就是 $K_j=\{t_i \mid f(t_i)=K_j, 1 \leqslant i \leqslant n, t_i \in D\}$。

聚类算法通常可分为以下几类：分层聚类法、基于质心的聚类法、基于密度的聚类法以及基于分布的聚类法。分层聚类法非常适用于分层数据；基于质心的聚类法，其聚类由中心矢量表示，但该矢量不一定是数据集的成员；基于密度的聚类法是基于某个阈值内的邻域密度，并允许任意形状的分布，包括球形聚类；当知道数据中的分布类型时，则可以选择基于分布的聚类法。对于特定的问题，最合适的聚类分析法通常需要通过实验来选择，除非有数学原因偏爱一种聚类模型而不是另一种聚类模型。为一种模型设计的算法通常会在包含完全不同模型的数据集上失败。

本章我们将探讨三种类型的聚类算法：层次聚类法、K 均值算法（基于质心的聚类法）和 DBSCAN（基于密度的聚类法）。它们是使用广泛的聚类算法，为后来出现的许多聚类算法提供了灵感。

## 3.1 Introduction

Clustering is similar to classification techniques but the classes/groups are not predefined. The groups are called clusters or viewed as special types of classification or segmentation. It is a type of unsupervised learning method and the task is to divide the population or datasets into a number of groups such that data points are more similar to each other in the same group than to those in other groups. It is a collection of objects based on *similarity* and *dissimilarity* among them. A simple example of customer segmentation can be found in EXAMPLE 3-1.

 EXAMPLE 3 – 1

> Suppose you are managing a rental store and wish to understand the preferences of your customers to scale up your business. Is it possible for you to look at the shopping details of each customer and devise a specific promotion strategy for every one of them? Absolutely not when the number of customers is very large. But, what you can do is to cluster all of your customers into, say, 10 groups based on their purchasing habits and personal attributes, then apply a separate strategy for costumers in each of these 10 groups. This is How the clustering technique is applied.

The cluster analysis is a general tool for gaining insights into the distribution of data to observe the characteristics of each cluster. Clustering can help in many fields such as biology. Clustering is useful to identify customers of a certain customer's record with similar conduct. Clustering can also help advertisers in their customer base to find different groups and their customer groups can be defined by buying patterns. In biology, it is used for the determination of plant and animal taxonomies for the categorization of genes with similar functionality and for insight into population-inherent structures. In a geographical observation database, clustering also makes it easier to find areas of similar use in the land. It helps to identify groups of houses and apartments by type, value and destination of houses. The clustering of documents on the web is also helpful for the discovery of hidden information. There are many other applications of clustering including recommendation engines, social network analysis, search result grouping and anomaly detection.

Some basic characteristics of clustering can be summarized in contrast to classification techniques:

- The number of clusters is not known.
- There may not be much apriori expertise related to the clusters. We have only one set of input data (without being labeled), and we do not previously know what the output will be.
- Cluster results are dynamic, i.e., users can switch from one cluster to another, as their preferences change.

The clustering problem definition is stated as: given a database of tuples (namely items or records) $D = \{t_1, t_2, \ldots, t_n\}$ for each $t_i = \{t_{i1}, t_{i2}, \ldots, t_{ip}\}$, where $p$ is the dimension of attributes, and the number of clusters to be generated is $k$, then the **clustering problem** is to define a mapping function $f : D \rightarrow \{1, 2, \ldots, k\}$, where each $t_i$ is assigned to one class $K_j$, $1 \leqslant j \leqslant k$; that is, $K_j = \{t_i \mid f(t_i) = K_j, 1 \leqslant i \leqslant n, \text{ and } t_i \in D\}$.

There are possibly over 100 published clustering algorithms in the published literature, a relative exhaustive summary of clustering approaches can be found in. Generally, the major prominent clustering algorithms can be classified into the following categories, which are shown in FIGURE 3-1. Hierarchical clustering can create a tree of clusters, it is well suited to hierarchical data, such as taxonomies. For centroid-based clustering, clusters are represented by a central vector, which may not necessarily be a member of the dataset. Density-based clustering is based on *density* in the neighborhood within some threshold and allows for arbitrary-shaped distributions including spherical-shaped clusters. When you know the type of distribution in your data, the distribution-based clustering is preferred which assumes that data is composed of distributions such as Gaussian distributions.

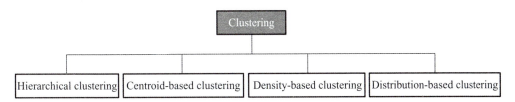

FIGURE 3-1　Categorization of clustering algorithms

As to the selection of clustering algorithms, please note that there is no objectively "correct" clustering algorithm, "clustering is in the eye of the beholder". The most appropriate clustering algorithm for a particular problem often needs to be chosen experimentally, unless there is a mathematical reason to prefer one cluster model over another. An algorithm that is designed for one kind of model will generally fail on a dataset that contains a radically different model.

In the following sections of this chapter, we will explore the first three types of clustering algorithms, specifically focusing on the Agglomerative Hierarchical Clustering, *K*-means algorithm (Centroid-based Clustering), DBSCAN (Density-based Spatial Clustering of Applications with Noise), which are the most widely used clustering algorithms. Sometimes they are considered as obsolete but they do provide inspirations for many later emerging clustering algorithms. Before moving to these algorithms we first discuss the similarity measures which are important to the performance of clustering algorithms.

## 3.2　Metrics Used for Clustering Analysis

### 3.2.1　Similarity metrics

Similarity is a metric that reflects the strength of the relationship between two data objects, which can be used as the measurement for clustering problems to evaluate the "alikeness" of the data objects within the same cluster. The similarity between any two data tuples $t_i$, $t_j \in D$ is defined as $sim(t_i, t_j)$, which is a mapping from $D \times D$ to the range of [0, 1]. Then we get the following inferences:

- $\forall t_i \in D$, $sim(t_i, t_i) = 1$.
- $\forall t_i, t_j \in D$, $sim(t_i, t_j) = 0$, if $t_i$ and $t_j$ are not alike at all.
- Given a cluster $K_j$, $\forall t_i, t_j \in K_j$ and $\forall t_m \notin K_j$, $sim(t_i, t_j) > sim(t_i, t_m)$.

There are some popular similarity measures, such as Dice similarity, Cosine similarity and Jaccard similarity.

- **Dice**:

$$sim(t_i, t_j) = \frac{2\sum_{h=1}^{p} t_{ih} t_{jh}}{\sum_{h=1}^{p} t_{ih}^2 + \sum_{h=1}^{p} t_{jh}^2} \tag{3.1}$$

- **Jaccard**:

$$sim(t_i, t_j) = \frac{\sum_{h=1}^{p} t_{ih} t_{jh}}{\sum_{h=1}^{p} t_{ih}^2 + \sum_{h=1}^{p} t_{jh}^2 - \sum_{h=1}^{p} t_{ih} t_{jh}} \tag{3.2}$$

- **Cosine**:

$$sim(t_i, t_j) = \frac{\sum_{h=1}^{p} t_{ih} t_{jh}}{\sqrt{\sum_{h=1}^{p} t_{ih}^2 \sum_{h=1}^{p} t_{jh}^2}} \tag{3.3}$$

- **Overlap**:

$$sim(t_i, t_j) = \frac{\sum_{h=1}^{p} t_{ih} t_{jh}}{\min\left(\sum_{h=1}^{p} t_{ih}^2, \sum_{h=1}^{p} t_{jh}^2\right)} \tag{3.4}$$

In these formulas, it is assumed that similarity is being evaluated between two vectors $t_i = \langle t_{i1}, t_{i2}, \ldots, t_{ik} \rangle$ and $t_j = \langle t_{j1}, t_{j2}, \ldots, t_{jk} \rangle$, and vector entries usually are assumed to be nonnegative numeric values.

### 3.2.2 Distance based metrics

A distance measure, $dis(t_i, t_j)$, instead of similarity, is often used in clustering analysis. Similar to the definition of similarity, for a cluster, $K_j$, $\forall t_i, t_j \in K_j$, and $\forall t_m \notin K_j$, $dis(t_i, t_j) < dis(t_i, t_m)$. The measure examines how "unlike" items are. Euclidean distance (or $L2$ norm) and Manhattan distance measures are two traditional metrics with the formulas shown below.

- **Euclidean**:

$$dis(t_i, t_j) = \sqrt{\sum_{h=1}^{p} (t_{ih} - t_{jh})^2} \tag{3.5}$$

- **Manhattan:**

$$dis(t_i, t_j) = \sum_{h=1}^{p} | t_{ih} - t_{jh} | \qquad (3.6)$$

To compensate for the different scales between different attribute values, the attribute values may be normalized to be in the range [0, 1]. Some clustering algorithms look only at numeric data, usually assuming metric data points. If nominal values rather than numeric values are used, some approach to determining the difference is needed. One method is to assign a difference of 0 if the values are identical and a difference of 1 if they are different. From the perspective of distance measures, a cluster can then be described by using several characteristic values, such as *centroid* ($C_j$), *radius* ($R_j$), and *diameter* ($D_j$) of the cluster for $K_j$ with $N$ points, the definitions are given as:

$$centroid = C_j = \frac{\sum_{i=1}^{N}(t_i)}{N}, t_i \in K_j \qquad (3.7)$$

$$radius = R_j = \sqrt{\frac{\sum_{i=1}^{N}(t_i - C_j)^2}{N}}, t_i \in K_j \qquad (3.8)$$

$$diameter = D_j = \sqrt{\frac{\sum_{i=1}^{N}\sum_{l=1}^{N}(t_i - t_l)^2}{N(N-1)}}, t_i, t_l \in K_j \qquad (3.9)$$

The *centroid* denotes the middle of a cluster; it may not be a sample point in the cluster. When a cluster center can not be represented by the centroid or when a centroid cannot be defined such as graphs, some clustering algorithms alternatively use the concept of *medoid* ($M_m$) to describe the central location of the cluster. Medoids are similar in concept to *centroids* or means, but medoids are always restricted to be members of the dataset. The *diameter* is the square root of the average mean squared distance between all pairs of points in the cluster, and the *radius* is the square root of the average mean squared distance from any point in the cluster to the *centroid*.

In fact, the hierarchical clustering method has a distinct advantage that any valid measure of distance can be used. There are many interpretations of the distance between clusters. Given clusters $K_i$ and $K_j$, one can calculate the distance between clusters by using several standard alternatives, such as:

- **Single link:** The minimum distance between each element in cluster $i$ and $j$. We thus have $dis(K_i, K_j) = \min(dis(t_i, t_j))$, $\forall t_i \in K_i$ and $\forall t_j \in K_j$.

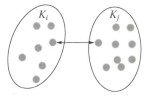

- **Complete link**: The maximum distance between an element in cluster $i$ and another element in cluster $j$, then we have $dis(K_i, K_j) = \max(dis(t_i, t_j))$, $\forall t_i \in K_i$ and $\forall t_j \in K_j$.

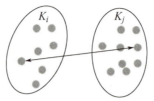

- **Average link**: The average distance between each element in cluster $i$ and $j$ is considered to be equal to: $dis(K_i, K_j) = \text{mean}(dis(t_i, t_j))$, $\forall t_i \in K_i$ and $\forall t_j \in K_j$.

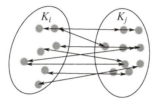

- **Centroid**: If clusters have a representative centroid, then the centroid distance is defined as the distance between the centroids of the clusters, i.e., $dis(K_i, K_j) = dis(C_i, C_j)$, where $C_i$ and $C_j$ are the centroids for $K_i$, $K_j$. Similarly, when using a *medoid* metric to represent a cluster, the distance between clusters can be defined as: $dis(K_i, K_j) = dis(M_i, M_j)$. These measures will be further illustrated in the cluster algorithms of hierarchical clustering and K-means algorithms.

## 3.3　Agglomerative Hierarchical Clustering Algorithm

The agglomerative algorithm is the most common type of hierarchical clustering method used to group objects in clusters based on their similarity/distance. The algorithm treats each data point as a single cluster at the beginning and then successively merges (or agglomerate) pairs of clusters until all clusters have been merged into one big cluster that contains all data points. For this reason, it is also called "bottom-up" hierarchical clustering algorithm. The result is a tree-based representation of the clusters, named *dendrogram*. In FIGURE 3-2, a traditional structure of dendrogram to represent clustering is given. As can be seen, the root contains one cluster where all elements are combined together. Internal nodes in the dendrogram represent new clusters formed by merging the clusters that appear as their children in the tree. The linkage criteria determine the metric used for the merge strategy (single, complete, or average linkage). All clusters created at a particular level are combined because the children clusters have a distance between them less than the distance threshold value associated with this level in the tree. Different agglomerative algorithms differ in how the clusters are merged at each level.

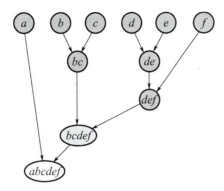

**FIGURE 3-2   A traditional structure of dendrogram to represent clustering**

The algorithm assumes the input dataset is a set of elements and pairwise distances between them are usually formatted with an $N \times N$ adjacent matrix $A$. The space required for $A$ is $\mathcal{O}(n^2)$ where there are $n$ items to cluster. Due to the iterative nature of the algorithm, matrix $A$ must be called multiple times to calculate dissimilarity between sets of observations for deciding which clusters should be combined. This makes it too slow for even medium datasets and becomes a very severe issue for large databases.

### 3.3.1  Single-link clustering

The single-link clustering is based on grouping clusters in a bottom-up fashion. At each step, two clusters are merged if the minimum distance between any points from each cluster is less than or equal to the threshold distance being considered.

Given a $N \times N$ proximity matrix $A$, contains all distances $dis(i, j)$, the matrix will be symmetric and its diagonals will be 0. The clusterings are assigned sequential numbers $0, 1, \ldots, n-1$, indicating the order of the clustering and $Len(k)$ is the level of the $k$-th clustering. A cluster with sequence number $m$ is denoted $(m)$ and the proximity between clusters $(r)$ and $(s)$ is denoted $dis(i, j)$, where the clusters can be viewed as an agglomerated point. Then the single-link clustering process can be described as the following steps:

**Step 1:** Start with the initial clustering with $Len(0)=0$ and sequence number $m=0$.

**Step 2:** Find the closest pair of elements/clusters in the current clustering analysis, say pair $(r)$, $(s)$, according to $dis[(r), (s)] = \min dis[(i), (j)]$, where the minimum is over all pairs of elements/clusters in the current clustering analysis.

**Step 3:** Increase $m$ to $m=m+1$. Merge clusters $(r)$ and $(s)$ into a single cluster to form the next clustering $m$, set $Len(m)=dis[(r),(s)]$.

**Step 4:** Update the proximity matrix $A$, by removing entries corresponding to clusters $(r)$ and $(s)$ replacing them with the newly formed cluster $(r, s)$. The distance between $(r, s)$ and each of the old elements/clusters $(k)$ is defined as $dis[(r,s),(k)]=$

$\min\{dis[(k),(r)], dis[(k),(s)]\}$.

**Step 5:** If all members are in one big cluster, stop. Otherwise, go to **Step 2**.

For illustration, we give an example (EXAMPLE 3-2) to exhibit how the algorithm works step by step.

 **EXAMPLE 3-2**

### Single-linkage clustering

Suppose we have five elements and proximity matrix $A_0$ with pairwise distance between them, shown in TABLE 3-1.

TABLE 3-1　Sample data of proximity matrix $A_0$

| Element | a | b | c | d | e |
|---|---|---|---|---|---|
| a | 0.0 | 5.0 | 7.0 | 9.0 | 3.0 |
| b | 5.0 | 0.0 | 2.5 | 2.0 | 8.5 |
| c | 7.0 | 2.5 | 0.0 | 1.0 | 8.0 |
| d | 9.0 | 2.0 | 1.0 | 0.0 | 5.5 |
| e | 3.0 | 8.5 | 8.0 | 5.5 | 0.0 |

**Step 1:** $L(0)=0$, $m=0$, $m=0$, each element is viewed as a cluster.

**Step 2:** $A_0(c,d)=1.0$ is the lowest value in $A_0$, then we merge $c$ and $d$ as the new single cluster $(c,d)$. Set $m=1$, $Len(1)=dis[(c),(d)]=1.0$.

**Step 3:** Update $A_0$ to $A_1$. Let $u$ denote the node to which $c$ and $d$ are now merged $(c,d)$. The distance between the rest elements and $(c,d)$ can be calculated.

$A_1((c,d),a)=dis[(c,d),(a)]=\min\{dis[(c),(a)],dis[(d),(a)]\}=\min(7.0,9.0)=7.0$
$A_1((c,d),b)=dis[(c,d),(b)]=\min\{dis[(c),(b)],dis[(d),(b)]\}=\min(2.5,2.0)=2.0$
$A_1((c,d),e)=dis[(c,d),(e)]=\min\{dis[(c),(e)],dis[(d),(e)]\}=\min(8.0,5.5)=5.5$

Then we get the new matrix $A_1$ below.

| Element | a | b | cd | e |
|---|---|---|---|---|
| a | 0.0 | 5.0 | 7.0 | 3.0 |
| b | 5.0 | 0.0 | 2.0 | 8.5 |
| cd | 7.0 | 2.0 | 0.0 | 5.5 |
| e | 3.0 | 8.5 | 5.5 | 0.0 |

**Step 4:** Now $A_1(b, cd) = 2.0$ is the lowest value in $A_1$, then we merge $cd$ and $b$ as the new single cluster $(b, cd)$. Set $m=2$, $Len(2) = dis[(b), (cd)] = 2.0$.

**Step 5:** Update $A_1$ to $A_2$. Let $v$ denote the node to which $b$ and $cd$ are now merged $(b, cd)$. The distance between the rest elements and $(b, cd)$ can be calculated.

$A_2((b,cd), a) = dis[b,cd,(a)] = \min\{dis[(b),(a)], dis[(cd),(a)]\} = \min(5.0, 7.0) = 5.0$

$A_2((b,cd), e) = dis[b,cd,(e)] = \min\{dis[(b),(e)], dis[(cd),(e)]\} = \min(8.5, 5.5) = 5.5$

Then we get the new matrix $A_2$ below.

| Element | a | bcd | e |
|---|---|---|---|
| a | 0.0 | 5.0 | 3.0 |
| bcd | 5.0 | 0.0 | 5.5 |
| e | 3.0 | 5.5 | 0.0 |

**Step 6:** Now $A_2(a, e) = 3.0$ is the lowest value in $A_2$ below, then we merge $a$ and $e$ as the new single cluster $(a, e)$. Set $m=3$, $Len(3) = dis[(a),(e)] = 3.0$.

**Step 7:** Update $A_2$ to $A_3$. Let $t$ denote the node to which $a$ and $e$ are now merged $(a, e)$. The distance between the rest elements and $(a, e)$ can be calculated.

$A_3((b,cd), (a,e)) = dis[(b,cd), (a,e)]$
$= \min\{dis[(bcd),(a)]\}, dis[(bcd),(e)] = \min(5.0, 5.5) = 5.0$

Then we get the new matrix $A_3$ below.

| Element | ae | bcd |
|---|---|---|
| ae | 0.0 | 5.0 |
| bcd | 5.0 | 0.0 |

**Step 8:** Now $A_3(ae, bcd) = 5.0$ is the lowest value in $A_3$, then we merge them as the final big single cluster $(ae, bcd)$. Set $m=4$, $Len(4) = dis[(ae), (bcd)] = 5.0$ and stop the iteration.

Based on the above results, we can draw the single-linkage dendrogram as FIGURE 3-3.

We can implement this in **Python** and the simple coding script is given as follows.

1. `import numpy as np`
2. `from scipy.cluster.hierarchy`
3. `import dendrogram, linkage`
4. `from scipy.spatial.distance import squareform`
5. `import matplotlib.pyplot as plt`
6. `# skipping the distance calculation part and directly using the Distance Matrix`
7. `matr= np.array([[0,5,7,9,3], [5,0,2.5,2.0,8.5], [7.0,2.5,0,1,8], [9,2,1,0,5.5], [3,8.5,8,5.5,0]])`

```
    8. dists= squareform(matr)
    9. # This step is where we mention its "Single Link" Clusterlinkage_ matrix =
linkage (dists, "single")
    10. dendrogram (linkage_ matrix, labels= ["a","b","c","d","e"])
    11. plt.title ("Single Link")
    12. plt.show ()
```

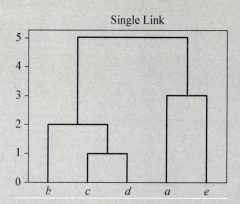

**FIGURE 3-3** The single-linkage dendrogram

Note that although the single linkage approach is simple, it suffers from tackling clustering analysis with long chains. Two clusters are merged if only any two points of them are close to each other. There may be points in the respective clusters to be merged that are far apart, but this does not influence the algorithm. So the final clusters may include points that are not related to each other at all but just simply happen to be close to each other.

### 3.3.2 Complete-link clustering

The complete-link clustering is another algorithm of agglomerative hierarchical clustering, which has similar steps to the single link algorithm. In this method, the distance between clusters equals the distance between those two points (one in each cluster) that are farthest away from each other. At each step, the two clusters separated by the shortest distance are combined. Following the same predefinition of mathematical symbols in single-link clustering, the detailed procedure of the algorithm consists of the following steps.

**Step 1:** Start with initial clustering with $Len(0)=0$ and sequence number $m=0$.

**Step 2:** Find the closest pair of elements/clusters in the current clustering analysis, say pair $(r)$, $(s)$, according to $dis[(r),(s)]=\min dis[(i),(j)]$, where the minimum is over all pairs of elements/clusters in the current clustering analysis.

**Step 3:** Increase $m$ to $m=m+1$. Merge clusters $(r)$ and $(s)$ into a single cluster to form the next clustering $m$, set $Len(m)=dis[(r),(s)]$.

**Step 4:** Update the proximity matrix, $A$, by removing entries corresponding to clusters $(r)$ and $(s)$ and replacing with them the newly formed cluster $(r, s)$. The distance between $(r, s)$

and each of the old elements/clusters $(k)$ is defined as $dis[(r,s),(k)]=\max\{dis[(k),(r)], dis[(k),(s)]\}$.

**Step 5:** If all members are in one big cluster, stop. Otherwise, go to Step 2.

For illustration, we again provide an example (EXAMPLE 3-3) to exhibit how the algorithm works step by step, just as what has been presented in the section of complete-linkage clustering.

**EXAMPLE 3-3**

**Complete-linkage clustering**

Suppose we have five elements and proximity matrix $A_0$ with pairwise distances between them, shown in TABLE 3-1.

**Step 1:** $L(0)=0$, $m=0$, each element is viewed as a cluster.

**Step 2:** $A_0(c, d)=1.0$ is the lowest value in $A_0$, then we merge $c$ and $d$ as the new single cluster $(c, d)$. Set $m=1$, $Len(1)=dis[(c),(d)]=1.0$.

**Step 3:** Update $A_0$ to $A_1$. Let $u$ denote the node to which $c$ and $d$ are now merged $(c, d)$. The distance between the rest elements and $(c, d)$ can be calculated.

$A_1[(c,d), a]=dis[(c,d),(a)]=\max\{dis[(c),(a)], dis[(d),(a)]\}=\max(7.0, 9.0)=9.0$

$A_1[(c,d), b]=dis[(c,d),(b)]=\max\{dis[(c),(b)], dis[(d),(b)]\}=\max(2.5, 2.0)=2.5$

$A_1[(c,d), e]=dis[(c,d),(e)]=\max\{dis[(c),(e)], dis[(d),(e)]\}=\max(8.0, 5.5)=8.0$

Then we get the new matrix $A_1$ below.

| Element | a | b | cd | e |
|---|---|---|---|---|
| a | 0.0 | 5.0 | 9.0 | 3.0 |
| b | 5.0 | 0.0 | 2.5 | 8.5 |
| cd | 9.0 | 2.5 | 0.0 | 8.0 |
| e | 3.0 | 8.5 | 8.0 | 0.0 |

**Step 4:** Now reiterate the above steps, we find $A_1(b, cd)=2.5$ is the lowest value in $A_1$, then we merge $cd$ and $b$ as the new single cluster $(b, cd)$. Set $m=2$, $Len(2)=dis[(b),(cd)]=2.5$.

**Step 5:** Update $A_1$ to $A_2$. Let $v$ denote the node to which $b$ and $cd$ are now merged $(b, cd)$. The distance between the rest elements and $(b, cd)$ can be calculated.

$A_2[(b,cd),a] = dis[(b,cd),(a)] = \max\{dis[(b),(a)], dis[(cd),(a)]\} = \max(5.0, 9.0) = 9.0$

$A_2[(b,cd),e] = dis[(b,cd),(e)] = \max\{dis[(b),(e)], dis[(cd),(e)]\} = \max(8.5, 8.0) = 8.5$

Then we get the new matrix $A_2$ below.

| Element | a | bcd | e |
| --- | --- | --- | --- |
| a | 0.0 | 9.0 | 3.0 |
| bcd | 9.0 | 0.0 | 8.5 |
| e | 3.0 | 8.5 | 0.0 |

**Step 6:** Now $A_2(a,e) = 3.0$ is the lowest value in $A_2$, then we merge $a$ and $e$ as the new single cluster $(a, e)$. Set $m=3$, $Len(3) = dis[(a),(e)] = 3.0$.

**Step 7:** Update $A_2$ to $A_3$. Let $t$ denote the node to which $a$ and $e$ are now merged $(a, e)$. The distance between the rest elements and $(a, e)$ can be calculated:

$A_3[(b,cd),(a,e)] = dis[(b,cd),(a,e)]$
$\qquad = \max\{dis[(bcd),(a)], dis[(bcd),(e)]\} = \max(9.0, 8.5) = 9.0$

Then we get the new matrix $A_3$ below.

| Element | ae | bcd |
| --- | --- | --- |
| ae | 0.0 | 9.0 |
| bcd | 9.0 | 0.0 |

**Step 8:** Now $A_3(ae, bcd) = 9.0$ is the lowest value in $A_3$, then we merge them as the final big single cluster $(ae, bcd)$, Set $m=4$, $Len(4) = dis[(ae), (bcd)] = 9.0$. Stop the iteration.

Based on the above results, we can draw the complete-linkage clustering as FIGURE 3-4.

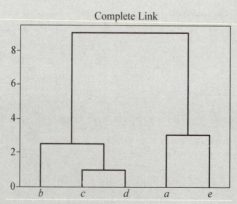

FIGURE 3-4  The complete-linkage clustering in EXAMPLE 3-3

We can implement this in **Python** and the simple coding script is given as follows.

```
1. import numpy as np
2. from scipy.cluster.hierarchy import dendrogram, linkage
3. from scipy.spatial.distance import squareform
4. import matplotlib.pyplot as plt
5. # skipping the distance calculation part and directly using the Distance Matrix
6. matr= np.array([[0,5,7,9,3], [5,0,2.5,2.0,8.5], [7.0,2.5,0,1,8], [9,2,1,0,5.5], [3,8.5,8,5.5,0] ])
7. dists= squareform(matr)
8. # This step is where we mention its "Single Link" Cluster
9. # linkage_matrix = linkage (dists, "single")
10. linkage_matrix = linkage (dists, "complete")
11. dendrogram (linkage_matrix, labels= ["a","b","c","d","e"])
12. plt.title ("Complete Link")
13. plt.show ()
```

Complete-linkage clustering can avoid the so-called chaining phenomenon when using the single linkage clustering. And the clusters tend to be more compact than those found via the single link technique. In fact, implementing a different linkage is simply a matter of using different metrics to calculate inter-cluster distances in the steps of updating the proximity matrix of the above algorithms. So it is hard to say one algorithm is guaranteed to be better than others, which depends on the problems considered, data structure, and the measurement scale.

### 3.3.3 Average-link clustering

There are two types of average linkage clustering algorithms, i.e., Weighted Pair Group Method with Arithmetic Mean (WPGMA) and Unweighted Pair Group Method with Arithmetic Mean (UPGMA). They are generally attributed to Sokal and Michener. The average link approach merges two elements/clusters based on the nearest average distance metrics. The mean distance between elements of each cluster used in UPGMA is defined as: $\frac{1}{|r|\cdot|s|}\sum_{i\in r}\sum_{j\in s}d(i,j)$. That is to say, at each clustering step, the updated distance between the merged clusters $r\cup s$ and the other cluster $X$ is calculated by the proportional averaging of the $d_{r,X}$ and $d_{s,X}$ distance:

$$d_{(r\cup s),X}=\frac{|r|\cdot d_{r,X}+|s|\cdot d_{s,X}}{|r|+|s|}$$

where $|r|$ and $|s|$ are the cardinality of the clusters $r$ and $s$. In the case of WPGMA, the distance is just the arithmetic mean of the average distances between $d_{r,X}$ and $d_{s,X}$.

EXAMPLE 3-4 illustrates the concrete process of clustering analysis using UPGMA.

EXAMPLE 3-4

**Average linkage clustering**

Suppose we have five elements and proximity matrix $A_0$ with pairwise distances between them, shown in the follow table.

| Element | a | b | c | d | e |
|---|---|---|---|---|---|
| a | 0.0 | 5.0 | 7.0 | 9.0 | 3.0 |
| b | 5.0 | 0.0 | 2.5 | 2.0 | 8.5 |
| c | 7.0 | 2.5 | 0.0 | 1.0 | 8.0 |
| d | 9.0 | 2.0 | 1.0 | 0.0 | 5.5 |
| e | 3.0 | 8.5 | 8.0 | 5.5 | 0.0 |

**Step 1:** $L(0)=0$, $m=0$, each element is viewed as a cluster.

**Step 2:** $A_0(c,d)=1.0$ is the lowest value in $A_0$, then we merge $c$ and $d$ as the new single cluster $(c,d)$. Set $m=1$, $Len(1)=dis[(c),(d)]=1.0$.

**Step 3:** Update $A_0$ to $A_1$. The distance between the rest elements and $(c,d)$ can be calculated.

$$A_1[(c,d),a]=dis[(c,d),(a)]$$
$$=\{dis[(c),(a)]\times1+dis[(d),(a)]\times1\}/(1+1)$$
$$=\{7.0\times1+9.0\times1\}/2=8.0$$
$$A_1[(c,d),b]=dis[(c,d),(b)]$$
$$=\{dis[(c),(b)]\times1+dis[(d),(b)\times1]\}/(1+1)$$
$$=\{2.5\times1+2.0\times1\}/2=2.25$$
$$A_1((c,d),e)=dis[(c,d),(e)]$$
$$=\{dis[(c),(e)]\times1+dis[(d),(e)]\times1\}/(1+1)$$
$$=\{8.0\times1+5.5\times1\}/2=6.75$$

Then we get the new matrix $A_1$ below.

| Element | a | b | cd | e |
|---|---|---|---|---|
| a | 0.0 | 5.0 | 8.0 | 3.0 |
| b | 5.0 | 0.0 | 2.25 | 8.5 |
| cd | 8.0 | 2.25 | 0.0 | 6.75 |
| e | 3.0 | 8.5 | 6.75 | 0.0 |

**Step 4:** Now reiterate the above steps, we find $A_1(b,cd)=2.25$ is the lowest value in $A_1$, then we merge $cd$ and $b$ as the new single cluster $(b,cd)$. Set $m=2$, $Len(2)=dis[(b),(cd)]=2.25$.

**Step 5:** Update $A_1$ to $A_2$. The distance between the remaining elements and $(b,cd)$ can be calculated.

$A_2[(b, cd), a] = dis[(b,cd),(a)] = \{dis[(b),(a)] \times 1 + dis[(cd),(a)] \times 2\}/(1+2) = \{5.0 \times 1 + 8.0 \times 2\}/3 = 7.0$

$A_2[(b, cd), e] = dis[(b,cd),(e)] = \{dis[(b),(e)] \times 1 + dis[(cd),(e)] \times 2\}/(1+2) = \{8.5 \times 1 + 6.75 \times 2\}/3 = 7.3$

Then we get the new matrix $A_2$ below.

| Element | a | bcd | e |
|---|---|---|---|
| a | 0.0 | 7.0 | 3.0 |
| bcd | 7.0 | 0.0 | 7.3 |
| e | 3.0 | 7.3 | 0.0 |

**Step 6:** Now $A_2(a, e) = 3.0$ is the lowest value in $A_2$, then we merge $a$ and $e$ as the new single cluster $(a, e)$. Set $m=3$, $Len(3) = dis[(a),(e)] = 3.0$.

**Step 7:** Update $A_2$ to $A_3$. Let $t$ denote the node to which $a$ and $e$ are now merged $(a, e)$. The distance between the rest elements and $(a, e)$ can be calculated.

$A_3[(a,e),(b,cd)] = dis[(a,e)(b,cd)]$
$= \{dis[(a),(bcd)] \times 1 + dis[(e),(bcd)] \times 1\}/(1+1)$
$= \{7.0 \times 1 + 7.3 \times 1\}/2 = 7.15$

Then we get the new matrix $A_3$ below.

| Element | ae | bcd |
|---|---|---|
| ae | 0.0 | 7.15 |
| bcd | 7.15 | 0.0 |

**Step 8:** Now $A_3(ae, bcd) = 7.15$ is the lowest value in $A_3$, then we merge them as the final big single cluster $(ae, bcd)$. Set $m=4$, $Len(4) = dis[(ae),(bcd)] = 7.15$. Stop the iteration.

Based on the above results, we can draw the single-linkage dendrogram as FIGURE 3-5.

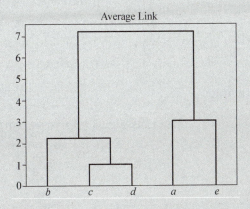

**FIGURE 3-5** The single-linkage dendrogram

## 3.4 K-means Algorithm

K-means clustering algorithms may be the most widely-used centroid-based clustering algorithm that aims to partition $n$ observations into $k$ clusters where each observation belongs to the cluster with the nearest mean serving as a prototype of the cluster. Since it involves assigning observations to clusters by minimizing the variance within each cluster or within-cluster sum-of-squares, it can be called a variance-minimizing approach and similar to Ward's minimum variance criterion used in the agglomerative hierarchical approach. The less variation we have within clusters, the more homogeneous (similar) the data points are within the same cluster.

The desired number of clusters, $k$, is specified as an input parameter. For a dataset $D = \{t_1, t_2, ..., t_n\}$ where $t_i$ is multidimensional. The way of K-means algorithm works is as the following steps.

**Step 1:** Randomly produce $k$ means (either being chosen from the dataset $D$ or just $k$ random values which may not be elements of $D$) for the centroids of the $k$ clusters, denoted by $m_i$, $1 \leqslant i \leqslant k$.

**Step 2:** Assign each data point to the cluster with the nearest centroid based on the squared Euclidean distance. Formally, the set of data point assigned to the $i^{th}$ cluster centroid be: $S_i = \{t_p : ||t_p - m_i||^2 \leqslant ||t_p - m_j||^2\}$, $\forall i$, where each $t_p$ is assigned to exactly one, even if it could be assigned to two or more of them.

**Step 3:** Update Centroids (means) in this step by taking the mean of all data points assigned to that cluster.

$$m_i = \frac{1}{|S_i|} \sum_{t_p \in S_i} t_p$$

**Step 4:** The difference between the old and the new centroids ($\Delta m_i$) is computed and the algorithm repeats these **Step 2** and **Step 3** until the difference is less than a threshold, say a very number. In other words, it repeats until the centroids do not move significantly.

The results of the algorithm include:
- The centroids of the $k$ clusters, which can be used to label new data.
- Labels for the training data (each data point is assigned to a single cluster).

### 3.4.1 A simple example

Now let us illustrate how the K-means algorithm works with a handcrafted example before implementing the algorithm in Scikit-Learn in Python, see EXAMPLE 3-5.

**EXAMPLE 3-5**

Suppose you are given the following two dimensional dataset to divide it into two clusters $C_1$ and $C_2$:

$D = \{(7, 8), (11, 14), (12, 13), (13, 16), (14, 15), (12, 13), (6, 4), (7, 5), (5, 8), (5, 6)\}$.

Based on the K-means algorithms introduce above, we at first randomly initialize the values for $C_1$ and $C_2$, here we use the values of the first two datasets, namely, (7, 8) and (11, 14). Secondly, we start looping between **Step 2** and **Step 3**. Then the Euclidean distance of each point to the two clusters is calculated as follow.

| Serial Number | Data Points | Squared Euclidean Distance From $C_1 = (7, 8)$ | Squared Euclidean Distance From $C_2 = (11, 14)$ | Assigned Cluster |
|---|---|---|---|---|
| A | (7, 8) | 0 | 52 | $C_1$ |
| B | (11, 14) | 52 | 0 | $C_2$ |
| C | (12, 13) | 50 | 2 | $C_2$ |
| D | (13, 16) | 100 | 8 | $C_2$ |
| E | (14, 15) | 98 | 10 | $C_2$ |
| F | (15, 13) | 50 | 2 | $C_2$ |
| G | (6, 4) | 17 | 125 | $C_1$ |
| H | (7, 5) | 9 | 97 | $C_1$ |
| I | (5, 8) | 4 | 72 | $C_1$ |
| J | (5, 6) | 8 | 100 | $C_1$ |

The centroids are updated to new values which are calculated by averaging the coordinates of the data points that belong to the same clusters. For cluster $C_1$, there is currently five points $\{A, G, H, I, J\}$, therefore the mean of the coordinates is $[(7+6+7+5+5)/5, (8+4+5+8+6)/5] = (6, 6.2)$, which is the renewed $C_1$ centroid. Similarly, for $C_2$, the coordinate mean of the other five points $\{B, C, D, E, F\}$ becomes $[(11+12+13+14+15)/5, (14+13+16+15+13)/5] = (13, 14.2)$.

For the next iteration, the new centroid values for $C_1(6, 6.2)$ and $C_2(13, 14.2)$ will be used and the whole process will be repeated. The iterations stop until the centroids in the last two steps converge to the same values. The next iterations are as follow.

| Serial Number | Data Points | Squared Euclidean Distance From $C_1 = (6, 6.2)$ | Squared Euclidean Distance From $C_2 = (13, 14.2)$ | Assigned Cluster |
|---|---|---|---|---|
| A | (7, 8) | 4.24 | 67.6 | $C_1$ |
| B | (11, 14) | 85.84 | 2 | $C_2$ |

continued

| Serial Number | Data Points | Squared Euclidean Distance From $C_1 = (6, 6.2)$ | Squared Euclidean Distance From $C_2 = (13, 14.2)$ | Assigned Cluster |
|---|---|---|---|---|
| $C$ | (12, 13) | 82.24 | 1.6 | $C_2$ |
| $D$ | (13, 16) | 145.04 | 3.6 | $C_2$ |
| $E$ | (14, 15) | 141.44 | 3.2 | $C_2$ |
| $F$ | (15, 13) | 127.24 | 8.2 | $C_2$ |
| $G$ | (6, 4) | 4.84 | 145 | $C_1$ |
| $H$ | (7, 5) | 2.44 | 113.8 | $C_1$ |
| $I$ | (5, 8) | 4.24 | 93.2 | $C_1$ |
| $J$ | (5, 6) | 1.04 | 122 | $C_1$ |

After this iteration, the values of $C_1$ and $C_2$ are the same as that they are at the end of the first step. This means that data can not be clustered any further and the centroids for $C_1$ and $C_2$ finally are (6, 6.2) and (13, 14.2), respectively.

The answer is finally $S_1 = \{A, G, H, I, J\}$ and $S_2 = \{B, C, D, E, F\}$. For determining a new data point which cluster should be assigned to, the distance between the data point and the centroids of each clusters is calculated and assigned to the cluster whose centroid is nearest to the data point.

### 3.4.2 K-means clustering with scikit-learn

In this part, we will implement K-means clustering with scikit-learn package in Python.

At first, importing the required libraries:

```
1. import matplotlib.pyplot as plt
2. # matplotlib inline
3. import numpy as np
4. from sklearn.cluster import KMeans
```

The next is to prepare the data that we want to cluster. In our example, it is a NumPy array of 10 rows and 2 columns. The scikit-learn library can work with Numpy array type data inputs without requiring any preprocessing. You can also change the input data by using different data values.

```
1. X= np.array([[7,8], [11,14], [12,13], [13,16], [14,15], [15,13], [6,4], [7,5], [5,8], [5,6]])
2. X.shape # (10,2)
```

Before creating the clusters, we can plot these data points and check if we can eyeball any clusters. To do so, execute the following code.

1. plt.scatter(X[:,0],X[:,1], label= 'True Position')

Intuitively, from FIGURE 3 - 6, we should divide the above data points into two clusters: by making a cluster of five points on the bottom left and one cluster of five points on the top right.

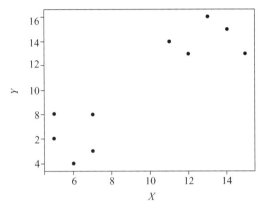

**FIGURE 3 - 6   The plot of data points in EXAMPLE 3 - 5**

It is now the moment to divide the data points into two clusters in Python. We can count on the following scripts.

1. kmeans = KMeans(n_clusters= 2)
2. kmeans.fit (X)
3. print (kmeans.cluster_centers_)

We then can obtain the centroid values that the algorithm generated for the final two clusters.

1. [[13. 0 14. 2]
2. [ 6. 0 6. 2]]

The values are the same as what we get manually for $C_1$ and $C_2$, the results of clustering are validated, and the algorithm performs well. Furthermore, we can see the labels of data points by executing the line.

1. print (kmeans.labels_)
2. [1 0 0 0 0 0 1 1 1 1]

Here 0 and 1 merely represent cluster IDs and have no mathematical meaning. If there were three clusters, the third cluster would have been represented by digit 2, for more clusters the cluster could be represented in a similar manner.

### 3.4.3  Choosing k

Evaluating the performance of a clustering algorithm is not as trivial as counting the number of errors or the precision and recall of a supervised classification algorithm. Contrary to supervised learning where we have the ground truth to evaluate the model's performance, clustering analysis doesn't have a solid evaluation metric that we can use to evaluate the outcome of different clustering algorithms.

To find the number of clusters for the dataset, you can run the K-means algorithm for a range of $k$ values and compare the results. And there are some other techniques for validating $k$, including cross-validation, information criteria, information-theoretic method, the Silhouette method, and the G-means algorithm. Moreover, estimating the distribution of data points across clusters provides insights into how the algorithm is splitting the data for each $k$. However, there is no right method for determining the exact value of $k$, sometimes expertise and background knowledge may help but usually in practice this is very hard to popularize to all kinds of problems.

In the following, we'll present two metrics, i.e., Elbow method and Silhouette analysis, which may provide us with some intuitions about how to choose the number of $k$.

Let us start with the Elbow Method, which consists of plotting the explained variation (Sum of Squared Error (SSE) between data points and their assigned centroids) as a function of the number of clusters, and picking the elbow of the curve as the number of clusters $k$ where SSE starts to flatten out. We will illustrate this process by using the dataset in EXAMPLE 3-5 and check where the curve might shape an elbow and flatten out. The code in Python is shown below and we can plot the results in FIGURE 3-7.

```
1. # Run the Kmeans algorithm and get the index of data points clusters
2. SSE = []
3. list_k = list(range(1, 10))
4. for k in list_k:
5.     km = KMeans(n_clusters= k)
6.     km.fit(X_std)
7.     sse.append(km.inertia_)
8. # Plot sse against k
9. plt.figure(figsize= (6, 6))
10. plt.plot(list_k, SSE, '- o')
11. plt.xlabel(r'Number of clusters (k)')
12. plt.ylabel('Sum of squared distance')
```

It can be seen that $k=2$ is an appropriate choice of the number of clusters. Please note that it's not always a good method to find out a proper number of clusters of the dataset in question because sometimes the curve is monotonically decreasing and might not have a distinctive point where the curve starts flattening out.

As to Silhouette analysis, it can be used to validate consistency within clusters of a dataset and measure how similar a data point is to its assigned cluster by contrast to other clusters. The value of Silhouette takes a range $[-1, 1]$. If one data point has a high value and is close to 1, this indicates the point is well matched to its own cluster and poorly

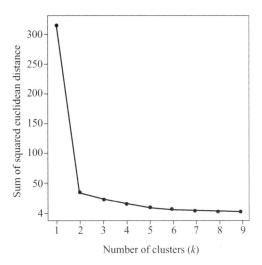

**FIGURE 3-7  Plot the results in EXAMPLE 3-5**

matched to neighboring clusters, then the clustering configuration is appropriate. When the value is $-1$, the data point is assigned to the wrong clusters. Therefore, we expect the value of Silhouette should be as large as possible and close to 1 to have a better result of clustering.

The definition of a Silhouette value of one sample data point $i$ is given as follows:

$$s(i) = \frac{b_i - a_i}{\max\{a_i, b_i\}}$$

in which, $a_i$ is be the mean distance between $i$ and all other data points in the same cluster, $b_i$ is the smallest mean distance of $i$ to all points in any other cluster, of which $i$ is not a member. A small value of $a_i$ implies the cluster is well-matched, and a bigger $b_i$ means that $i$ is badly matched to its neighboring cluster. Thus the mean $s(i)$ over all data points considered is a measure of how appropriately the data points have been clustered. We again use the dataset in EXAMPLE 3-5 to show how the Silhouette value be calculated in Python because it is obvious that there are most likely only two clusters of data points, i.e., $k=2$.

```
1. silhouette_mean = []
2. for i in range(2,9)
3. kmeans_fit = KMeans(n_clusters = i).fit(X)
4. silhouette_mean.append(silhouette_score(X, kmeans_fit.labels_))
5. plt.plot(range(2,9), silhouette_mean)
```

The silhouette value with different number of clusters as FIGURE 3-8 shows, the average Silhouette value has the highest value of 0.74 when the number of clusters is set to 2. With the increase in the number of clusters, the mean of the Silhouette value decreased dramatically. This indicates the number of clusters of two is a good cluster pick for the data points in EXAMPLE 3-5.

In summary, K-means clustering algorithm scales well to large-scale samples and has been successfully applied in market segmentation, computer vision, and astronomy among

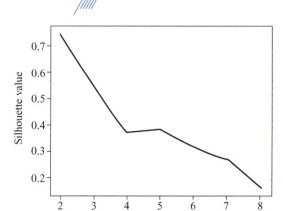

FIGURE 3-8  The silhouette value with different number of clusters

many other domains. Note that there are some differences between K-means and agglomerative hierarchical clustering. Firstly, agglomerative hierarchical clustering can't handle big data well but K-means clustering can. This is because the time complexity of K-means is linear i.e. $\mathcal{O}(n)$ while that of hierarchical clustering is quadratic $\mathcal{O}(n^2)$. Secondly, the number of clusters you want to obtain should be given with prior knowledge, and results produced by K-means clustering might differ for different running times since we start with random choices of clusters. While results are reproducible for hierarchical clustering algorithms and you can stop at whatever number of clusters you find appropriate in hierarchical clustering by interpreting the dendrogram. Thirdly, K-means works well when the shape of the clusters is hyperspherical (like a circle in 2D, a sphere in 3D). However, it suffers as the geometric shapes of clusters deviate from spherical shapes. Another point is that the algorithm has a loose relationship to the *K-nearest neighbor classifier*, a popular machine learning technique for classification that is often confused with K-means due to the name. To be a good practitioner, it's important to know the assumptions behind algorithms/methods so that you would have a pretty good idea about the strength and drawbacks of each algorithm. This will help you decide when to use what method under what circumstances.

## 3.5  DBSCAN Clustering Algorithm

DBSCAN, standing for "density-based spatial clustering of applications with noise", is based on the idea that for each point of a cluster, the neighborhood of a given radius has to contain at least a minimum number of points. DBSCAN can handle datasets of different arbitrary shapes that are hard for K-means and hierarchical clustering algorithms, which are only suitable for analyzing spherical-shaped clusters or convex clusters. Shapes of non-spherical dataset in FIGURE 3-9. It can also be very useful to identify outlier points that do not belong to any cluster. Then it is one of the most common clustering algorithms and also most cited in the scientific literature of data mining and machine learning.

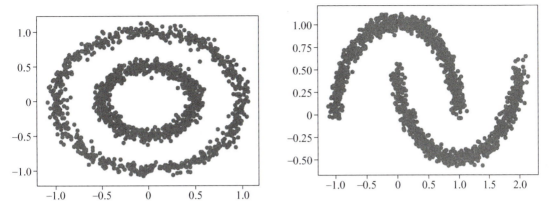

**FIGURE 3-9** **Shapes of non-spherical dataset**

To identify dense regions, DBSCAN generally requires two parameters.

- *eps*: the radius of neighborhood around a point $x$, which defines that if the distance (usually Euclidean distance) between two points is lower or equal to this threshold value (*eps*), these points are considered neighbors. Note that If the *eps* value is specified as very small, then this may lead to the most part of the dataset being labeled as outliers. While setting *eps* to be very large, the majority of the data points will be merged into a single and big cluster.
- *minPoints*: the minimum number of neighbor points within the *eps* of a given point $x$, and the point $x$ is marked as a *core point* if its neighbor's number is greater than or equal to *minPoints* which should be predefined as a priori. If the number of its neighbors within *eps* is less than *minPoints*, but it belongs to the neighborhood of the *core point*. A point that is neither a *core point* nor a *border point* is called *noise or outlier*, which has fewer neighbors than *minPoints* within the distance of *eps*.

FIGURE 3-10 shows the three types of points (core, border and noise points) setting *minPoints* = 6. We can see $A$ is a core point because the area surrounding it within *eps* contains 6 points (including $A$ itself). Point $B$ is a border point because its neighbors are less than 6, but it belongs to the *eps*-neighborhood of the core point $A$. Obviously, $C$ is a *noise* point. Two points $a$ and $b$ are *density-connected* if there is a point $c$ such that both $a$ and $b$ are reachable from $c$. Density-connectedness is symmetric. A cluster has two properties: (1) All points within the cluster are mutually density-connected; (2) If two points are density-connected, they belong to the same cluster.

Based on these definitions above, the algorithmic steps for DBSCAN clustering is outlined as follows.

**Step 1:** Arbitrarily pick up a point $x_i$ in the dataset, then compute the distance between $x_i$ and the other points. Find all neighbor points within *eps* of the starting point ($x_i$). If the neighbor count of it is greater than or equal to a predefined *minPoints*, then marked $x_i$ as a *core point* or visited.

**Step 2**: For each *core point* if it is not already assigned to a cluster, create a new cluster. Find recursively all its *density-connected points* and assign them to the same cluster as the *core point*.

**Step 3**: Iterate through the remaining unvisited points in the dataset. Assign each non-core point to the same cluster as a proximal *core point* belongs to, otherwise, assign it to noise.

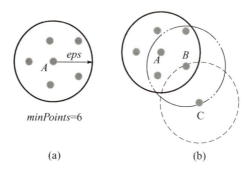

FIGURE 3 − 10　Three types of points in DBSCAN

### 3.5.1　Implementation of DBSCAN in Python using sklearn

We follow the demo example code from the introduction webpage of DBSCAN clustering algorithm, which is used to cluster a spherical dataset. The example first generates 750 spherical training data points with corresponding labels. After that standardize the features of your training data and at last, then apply DBSCAN from the sklearn library. FIGURE 3 − 11 illustrates the final results of three clusters, in which large circles indicate core points, smaller circles denote non-core point. These black points represent noises of the generated datasets.

```
1. import numpy as np
2. from sklearn.cluster import DBSCAN
3. from sklearn import metrics
4. from sklearn.datasets import make_blobs
5. from sklearn.preprocessing import StandardScaler
6. ###########################################
7. # Generate sample data
8. centers = [[1, 1], [-1,-1], [1,-1]]
9. X, labels_true = make_blobs(n_samples= 750, centers= centers, cluster_std=
   0.4, random_state= 0)
10. X = StandardScaler().fit_transform(X)
11. ###########################################
12. # Compute DBSCAN
```

```
13. db = DBSCAN(eps= 0.3, min_samples= 10).fit(X)
14. core_samples_mask = np.zeros_like(db.labels_, dtype= bool)
15. core_samples_mask[db.core_sample_indices_] = True
16. labels = db.labels_
17. # Number of clusters in labels, ignoring noise if present.
18. n_clusters_ = len(set(labels))- (1 if- 1 in labels else 0)
19. n_noise_ = list(labels).count(- 1)
20. print('Estimated number of clusters: % d' % n_clusters_)
21. print('Estimated number of noise points: % d' % n_noise_)
22. print("Silhouette Coefficient: % 0.3f" % metrics.silhouette_score(X, labels))
23. # # # # # # # # # # # # # # # # # # # # # # # # # # # # # # # # # # # # # # # #
24. # Plot result
25. import matplotlib.pyplot as plt
26. # Black removed and is used for noise instead.
27. unique_labels = set(labels)
28. colors = [plt.cm.Spectral(each)
29. for each in np.linspace(0, 1, len(unique_labels))]
30. for k, col in zip(unique_labels, colors):
31. if k = = - 1:
32. # Black used for noise.
33. col = [0, 0, 0, 1]
34. class_member_mask = (labels = = k)
35. xy = X[class_member_mask & core_samples_mask]
36. plt.plot(xy[:, 0], xy[:, 1], 'o', markerfacecolor= tuple(col),
37. markeredgecolor= 'k', markersize= 14)
38. xy = X[class_member_mask & ~ core_samples_mask]
39. plt.plot(xy[:, 0], xy[:, 1], 'o', markerfacecolor= tuple(col),
40. markeredgecolor= 'k', markersize= 6)
41. plt.title('Estimated number of clusters: % d' % n_clusters_)
42. plt.show()
```

As mentioned, DBSCAN is suitable for clustering analysis of non-spherical datasets. Next, we apply it to a dataset of circle shapes. The Python script is given as follows.

```
1. import numpy as np
2. import matplotlib.pyplot as plt
3. from sklearn import metrics
4. from sklearn.datasets import make_circles
5. from sklearn.preprocessing import StandardScaler
6. from sklearn.cluster import DBSCAN
7. X, y = make_circles(n_samples= 750, factor= 0.3, noise= 0.1)
8. X = StandardScaler().fit_transform(X)
9. y_pred = DBSCAN(eps= 0.3, min_samples= 10).fit_predict(X)
10. plt.scatter(X[:,0], X[:,1], c= y_pred)
```

```
11. print ('Number of clusters: {}'.format(len(set(y_pred[np.where(y_pred!=-1)]))))
12. print ('Number of outliers: {}'.format(len((y_pred[np.where(y_pred==-1)]))))
13. print ('Homogeneity: {}'.format(metrics.homogeneity_score(y, y_pred)))
14. print ('Completeness: {}'.format(metrics.completeness_score(y, y_pred)))
```

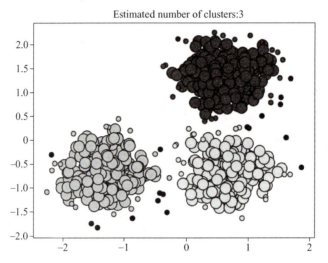

FIGURE 3-11  The final results of three clusters in DBSCAN

By running the above code, the clutering results is shown in FIGURE 3-12. It can be seen that the algorithm segment the non-linear dataset very well with two clusters, and the homogeneity score is 1.0 which indicates that all of the clusters contain only data points which are members of a single group.

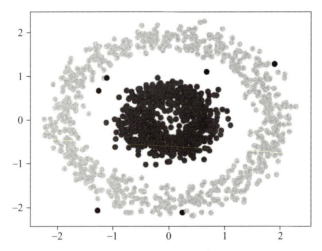

FIGURE 3-12  The clustering results of non-spherical data in DBSCAN

### 3.5.2  Rules of thumb for setting eps and minPoints

Before finishing this section of the introduction of density-based clustering algorithms, it should be pointed out although DBSCAN can learn clusters with arbitrary shapes, sometimes it is somewhat more arduous to tune as opposed to K-means. Parameters like the *eps* and *minPoints* for DBSCAN are less intuitive to reason about compared to the number of clusters parameter for K-means, so it would be difficult to choose good initial parameter values for the algorithm, in particular when the dataset and scale are not well understood by the practitioner. Nevertheless, there are some rules of thumb for the parameter estimations.

- Ideally, the value of *eps* is given by the problem to solve (e. g. a physical distance), and *minPoints* is then the expected minimum cluster size.
- A low *minPoints* will create clusters for outliers or noise. So we do not recommend to choose a too-small value for it. As a rule of thumb, *minPoints* can be derived from the number of dimensions (*dim*) in the dataset, usually, *minPoints* can be set as 2 times of *dim*, but it may be necessary to choose larger values for large datasets. Another heuristic approach is using $\ln(n)$, where $n$ is the sample size of the dataset to be clustered.
- For *eps*, you can choose the value by using a $k$-distance graph for your dataset and determine a "*knee*". The idea is to calculate the average of the distances of every point to its $k$ nearest neighbors. The $k$-distances are plotted in ascending order. The aim is to determine the "*knee*", which determines the optimal *eps* value. A *knee* indeed is a threshold where a sharp change occurs along the $k$-distance curve. Note that there's no guarantee that there will be a strong *knee* or even a bend at all in the graph which completely depends on the distribution of the dataset.

## EXERCISES

1. Show the dendrogram created by the single, complete, and average link clustering algorithms using the following adjacency matrix.

| Item | A | B | C | D |
|------|---|---|---|---|
| A | 0 | 7 | 4 | 5 |
| B | 7 | 0 | 2 | 6 |
| C | 4 | 2 | 0 | 3 |
| D | 5 | 6 | 3 | 0 |

2. Suppose you are given the following two-dimensional dataset, using K-means algorithm to divide it into three clusters:
$D=\{ (7,8),(11,14),(12,13),(13,16),(14,15),(12,13),(6,4),(7,5),(5,8),(5,6)\}$.

3. Given two tuples $A=(5,0,3,3,3,4)$ and $B=(4,1,4,6,2,3)$, how similar are $A$ and $B$? Please use metrics of dice similarity, cosine similarity, and Jaccard similarity and overlap similarity to calculate the similarity between the two vectors.

4. Use the DBSCAN algorithm to cluster the dataset in Exercise 2 with $eps=6$,

min $Points=3$. Is the result different from that of K-means?

5. (Research) Perform a literature review of recently developed clustering algorithms. Describe their approach, identify where they are applicable, and their performance.

本章配套资源

# Chapter 4
# Association Rules Mining

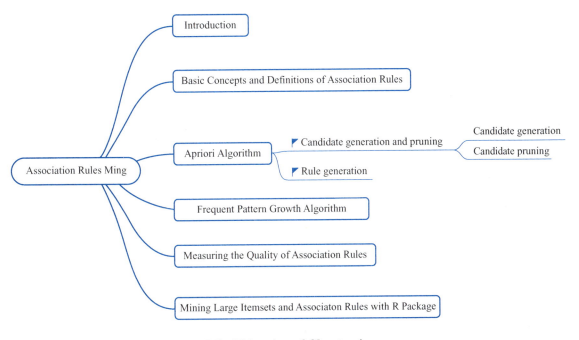

**Mind Mapping of Chapter 4**

关联规则挖掘技术经常用于发现隐藏在市场交易数据库或其他数据库中的项目集之间的有趣信息或关系。这些隐藏信息可以支持各种与业务相关的应用进程,包括客户行为分析、产品促销、目录设计和商店布局等。例如,在许多数据挖掘的解读中,发现尿布和啤酒之间的关系是一个非常有名和经典的例子:当一个男人去商店购买尿布时,也可能会购买啤酒。关联规则还用于许多其他领域,如医疗诊断、预测电信故障以及基于客户主观体验的推荐系统。

假设淘宝在线商店在给定的交易周期内有200000笔客户交易。其中,大约5000笔尿布销售记录(占总交易的2.5%),5500笔购买啤酒的记录(占总交易的2.75%),这其中又有大约4000笔记录包括购买尿布和啤酒(占总交易的2%)。根据这些百分比,我们可以计算出所有购买尿布的人中有80%也购买了啤酒,这代表了尿布和啤酒之间的关联规则。当应用关联规则挖掘时,商店所有者可以确定向客户交叉销售其产品的新机会。关联规则可以从数据库中的项集计算中得出,该项集通常由两部分组成:前置(IF)或左侧(LHS)以及后置

(THEN)或右侧（RHS），即 A=>B（IF A，THEN B）。前置部分是数据集中的项集，后置部分是与前置项集一起出现的项集。前置项和后置项都可以有多个项。我们将在以下部分提供正式定义。然而，应当指出不应将关联规则视为任何一种因果关系或相关性。因果关系是指原因与其结果之间或随时间发生的定期相关事件或现象之间的影响关系。

在本章我们将介绍关联规则挖掘技术的工作原理。首先，介绍关联规则的基本概念和定义；其次，用6个小节详细描述了一些先验算法（如Apriori算法），包括候选集生成、剪枝及规则生成等；再次，介绍了频繁增长模式算法；再次介绍了一些衡量关联规则质量的指标；最后，使用R语言包在"杂货"数据库上演示了频繁项集和关联规则挖掘过程。

## 4.1 Introduction

Association rules mining techniques are frequently used to discover interesting information/relationships hidden among sets of items in the market transaction databases or other data repositories. The uncovered information could support a variety of business-related applications such as customer behavior analysis, marketing promotions, catalog design and store layout. For instance, finding the relationship between diapers and beers is a very famous and classic example in many readings of data mining which claims that when a man going to a store to buy diapers may also possibly buy beers. Detailed information about this fictional story is illustrated in EXAMPLE 4-1. Association rules are also used in many other fields including medical diagnosis, predicting failures in telecommunications, and recommendation systems based on subjective experience of customers.

 EXAMPLE 4-1

Given that a Taobao online store has 200000 customer transactions for a given transaction period. Of which, about 5000 transactions (2.5% of total transactions) include the record of diapers sales. About 5500 transactions (5500/200000=2.75%) include the purchase of beers. Of those, about 4000 transactions (4000/200000=2%) include both the purchase of diapers and beers. Based on the percentages, we can calculate that 80% of all diaper purchases also buy beers, which represents an association rule between diapers and beers. It is useful for a store owner to identify new opportunities for cross-selling their products to customers when association rules mining is applied.

Association rules can be calculated from itemsets in a database, which usually consists of two parts: an antecedent (IF) or left-hand-side (LHS) and a consequent (THEN) or right-hand-side (RHS), denoting as A=>B (IF A, THEN B). The antecedent part is an itemset(s) from the dataset and the consequent part is an itemset(s) that appear(s) together with the antecedent itemset(s). Note that both antecedents and consequents can have multiple items. We will present a formal definition in the following parts. However, it should be pointed out that association rules should not be viewed as any kind of causality or correlation. Causality instead refers to the influential relationship between a cause and its effect or between regularly correlated events or phenomena occurring over time (e. g. , Raining leads to ground wet).

In this chapter, we will introduce how association rules mining technique works. Firstly, basic concepts and definitions are presented, then, some algorithms, particularly Apriori algorithms are described in detail. In the third section, we present some metrics on measuring the quality of association rules. Lastly, an R package named arules is used to mine large itemsets and association rules for demonstration.

## 4.2 Basic Concepts and Definitions of Association Rules

The association rules mining is to uncover how itemsets in a database are associated with each other. So the ways to describe the association is the first step for rules mining. Generally, two metrics to measure the strength of an association rule are used in many algorithms, i. e. , **Support** and **Confidence**. Throughout the chapter, we use the data in TABLE 4-1 to illustrate the basic concepts and related algorithms. Straightforward, there are 5 transactions ($t_1$ to $t_5$) and 5 items: {Bread, Jelly, PeanutButter, Milk, Beer}. Note that the example here is rarely small. In practice, a rule requires a **Support** count of hundreds of transactions before it can be regarded as statistically significant, and datasets in a supermarket actually often contain millions of transactions.

TABLE 4-1  Sample Data to Illustrate Association Rules

| Transaction | Item |
|---|---|
| $t_1$ | Bread, Jelly, PeanutButter |
| $t_2$ | Bread, PeanutButter |
| $t_3$ | Bread, Milk, PeanutButter |
| $t_4$ | Beer, Bread |
| $t_5$ | Beer, Milk |

More formally, given an itemset $I=\{I_1,I_2,\ldots,I_m\}$, and a database of market transactions $D=\{t_1,t_2,\ldots,t_n\}$ where $t_i=\{I_{i1},I_{i2},\ldots,I_{ik}\}$ and $I_{ij}\in I$, this means each transaction contains a subset of itemset $I$. If an itemset has $k$ items, it is called a $k$-itemset. The empty/null itemset does not contain any items. Then we define an **association rule** with the form $X\Rightarrow Y$ where $X$, $Y\subset I$ and $X\cap Y=\varnothing$. In applications, we usually do not care about all implication rules but only those are important and significant ones measured by **Support** and **Confidence** metrics.

**The Support (s)** for an association rule $X\Rightarrow Y$ is the proportion of transactions in the database that contain ($X\cap Y$). Mathematically, The Support formula can be written as: $Support(X\Rightarrow Y)=\dfrac{\sigma(X\cup Y)}{n}$, where $(X\cup Y)$ is the support count, denoting the number of transactions in the database containing $X\cup Y$. In TABLE 4-1, the $s$ of the rule {Bread}=>{PeanutButter} is 60% (=3/5), since the support count of {Bread, PeanutButter} is 3, and the total number of transactions is 5. Some items purchased by customers can have much larger impacts on the revenue of a business enterprise. These itemsets are identified as significant and important when the **Support** values are above a certain proportion. In other words, the **Support** metric is actually used to quantify if an itemset is large (frequent) or not.

The **Confidence or Strength ($\alpha$)** for an association rule $X \Rightarrow Y$ is the ratio of the number of transactions in the database that contain $X \cup Y$ to the number of transactions that contain $X$. Similarly, we can state the definition with mathematical expressions as: **Confidence**$(X \Rightarrow Y) = \frac{\sigma(X \cup Y)}{\sigma(X)}$. **Confidence** measures the rule strength and can be interpreted as an indication of how frequently the rule inferred to be true. The higher the **Confidence**, the more probably it is for RHS to co-occur with LHS for a given rule. For instance, the $\alpha$ of the rule {Bread}=>{PeanutButter} is 0.75 (=3/4), obtained by dividing the support count of {Bread, PeanutButter} (3) by the support count of {Bread} (4). This indicates that 75% of the time when *Bread* is purchased, the *PeanutButter* is also purchased by customers. The rule is stronger than {Jelly}=>{Milk} because no *Milk* is purchased when *Jelly* is bought based on the sample dataset. This can provide very useful information for the executive of a grocery store to promote their products to customers. Note that one weakness of the **Confidence** metric is that it might distort the importance of a rule. For the rule {Bread}>{PeanutButter}, it only accounts for how popular *Bread* is, but not *PeanutButter*. If *PeanutButter* is also very popular in general, there should be a higher probability that a transaction containing *Bread* will also contain *PeanutButter*, thus indues the increment of the confidence value. We will discuss this issue in a later section.

For most association rules mining algorithms, one important task is to find all the itemsets that satisfy the *minsup* threshold. These itemsets are called large (or frequent) itemsets. L is used to denote the complete set of large itemsets and $L$ to indicate a specific large itemset. Once large itemsets have been discovered, the next step is to generate all the high-confidence rules from these large itemsets. The rules $X \Rightarrow Y$ must have $X \cup Y$ in large itemsets. Although mining large itemsets is easier than rule generation, the efficiency of association rule algorithms usually differs in the complexity of finding large itemsets. Given an itemset $I$ of size $m$, there are $2^m$ subsets. Then the potential amount of large itemsets is then $2^m - 1$ ("1" is the empty set that generates meaningless rule), when $m=30$, the number becomes 1073741823 that is a seemly impossible task to complete. Many real transactional datasets are extremely large in terms of the volume of transactions as well as the size of itemsets. So, to generate the association rules is challenging and often involved in how to efficiently discover large itemsets. These potentially large itemsets are called *candidates*, and correspondingly, the set of all counted (potentially large) itemsets is the *candidate itemset* ($C$). The size of $C$ is an important performance measure used for many association rule algorithms.

Lastly, we give the definition of association rules as: for an itemset $I=\{I_1, I_2, \ldots, I_m\}$ and a database of market transactions $D=\{t_1, t_2, \ldots, t_n\}$ where $t_i = \{I_{i1}, I_{i2}, \ldots, I_{ik}\}$ and $I_{ij} \in I$, the association rule problem is to identify all association rules $X \Rightarrow Y$ with a minimum support and confidence threshold $(s, \alpha)$, which values are given as input to the algorithm.

## 4.3 Apriori Algorithm

There are many association rules mining algorithms proposed in the literature. As this is a textbook, it is not intended to present an exhaustive introduction of these algorithms. Here we limit our focus mainly on Apriori algorithm, which is the most famous association rule algorithm and widely used in much commercial software such as SPSS. The algorithm uses a simple *a prior* belief about the **large itemset property** to reduce the computational complexity of large itemset generation, that is, *any subset of a large itemset must also be large*. In other words, if an itemset meets the requirement of *minsup*, all of its subsets are *downward closed*.

To illustrate the Apriori principle, consider the itemset lattice shown in FIGURE 4-1. The nonempty subsets are seen as $\{AC, AD, CD, A, C, D\}$. If $\{A, C, D\}$ is large, so is each of its subsets. Moreover, recall the definition of *Support*, which indicates how frequently an itemset appears in a database. We know that if $\{A\}$ does not meet a given $s$, it is unnecessary to consider $\{A, C, D\}$ or even any itemset containing $\{A\}$ as a possible large itemset.

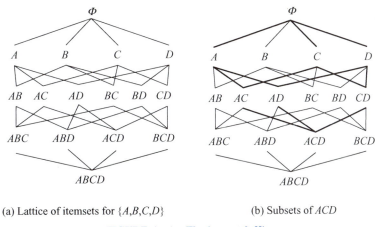

(a) Lattice of itemsets for $\{A,B,C,D\}$  (b) Subsets of $ACD$

**FIGURE 4-1 The itemset laffice**

Now let's see how the Apriori algorithm works. Given threshold values for *minsup* and *minconf*, the basic idea of Apriori algorithm is to generate candidate itemsets of a specific size and then scan the database to see if they are large. To put it simply, the actual process consists of two phases.

(1) Identifying all the itemsets that meet a *minsup* threshold to generate candidate itemsets and large itemsets.

(2) Creating rules from these large itemsets meets a *minconf* threshold.

For the first phase, there are multiple iterations involved in evaluating the support of a set of increasingly large itemsets. During scan $i$ candidates of size, $C_i$, are counted. Only those large candidates are used to generate candidates for the next iteration. $L_i$ are used to generate $C_{i+1}$. An itemset is considered as a candidate only if all its subsets also are large. In order to generate candidates of size $i+1$, joins are made for large itemsets with length $i$ found in the previous iteration. The algorithm terminates when no further successful extensions are found. In a sense, the Apriori is mainly used to generate large itemsets rather than generate association rules.

TABLE 4-2 shows results of applying Apriori algorithm with datasets in TABLE 4-1 with $s=30\%$ which means that the support count equals 1.5, and a *minconf* $\alpha=50\%$.

TABLE 4-2 Results of applying Apriori algorithm with dataset in TABLE 4-1

| Iteration | Candidates | Count | Large Itemsets |
|---|---|---|---|
| 1 | {Beer} | 2 | {Beer}, {Bread} {Milk}, {PeanutButter} |
| | {Bread} | 3 | |
| | {Jelly} | 1 | |
| | {Milk} | 2 | |
| | {PeanutButter} | 2 | |
| 2 | {Beer, Bread} | 1 | {Bread, PeanutButter} |
| | {Beer, Milk} | 1 | |
| | {Beer, PeanutButter} | 1 | |
| | {Bread, Milk} | 1 | |
| | {Bread, PeanutButter} | 2 | |
| | {Milk, PeanutButter} | 1 | |

Regarding the candidate 2-itemsets ($l_2$), for example, the pair {Beer, PeanutButter}, we can see its subsets, including {Beer} and {PeanutButter}, which are large 1-itemsets found in the first step. And the confidence of the rule {Bread}→{PeanutButter} is 2/3, which is greater than the *minconf* (0.5) denoting that it is a valid rule. In this example, there are no candidates of 3-itemsets because there is only one large itemset of size two. Note that, in practice, a business manager would typically be more interested in having a *complete* list of large itemsets with a length of two or above rather than single ones.

### 4.3.1 Candidate generation and pruning

(1) Candidate generation. $C_i$ candidate itemsets are generated by combining the $L_{i-1}$ with itself, which is found in the previous phrase. The pseudocode for candidate generation algorithm called candidate-gen is shown as follows.

---

**Input:**
　　$L_{i-1}$　　　　　　//Large itemsets of size $i-1$
**Output:**
　　$C_i$　　　　　　　//Candidate of length $i$

**Candidate-gen algorithm:**
$C_i = \emptyset$;
**for each** $I \in L_{i-1}$ **do**
  **for each** $J \neq I \in L_{i-1}$ do
    if $i-2$ of the elements in $I$ and $J$ are equal then
    $C_k = c_k \cup \{I \cup J\}$

---

A straightforward but arduous approach to generate candidate itemsets is to list each $k$-itemset as a possible candidate. Given $m$ items, the number of candidate itemsets of size $k$ is $C_m^k$. For this simple method, however, the overall complexity is rather high because it will impose massive computations for the pruning process.

There are many other alternative methods to generate candidate itemsets, such as $F_{k-1} \times F_1$ method. Each large $k$-itemset is composed of a large $(k-1)$-itemset and a large 1-itemset. $F_{k-1} \times F_{k-1}$ method that merges a pair of large $(k-1)$-itemset only if their first $k-2$ items are identical. For detailed information about the two methods, please see the reference.

(2) Candidate pruning. After the generation of candidate itemsets from a database, a pruning operation should be taken to drop off any candidates that have subsets with a length of $i-1$, which are not large. So it can effectively reduce the total number of candidate itemsets by using *support counting*. However, to update the support counts of each candidate itemset is computationally expensive because it requires comparing each transaction with each candidate itemset, in particular, when the total numbers of transactions and candidate itemsets are very huge. Support counting using a hash tree is a comparatively effective method without having to traverse all the candidate itemsets which would reduce the number of comparisons and the main-memory requirements for large transaction datasets. EXAMPLE 4-2 illustrates this method roughly.

**EXAMPLE 4-2**

Given 15 candidate itemsets of length 3: {1 4 5}, {1 2 4}, {4 5 7}, {1 2 5}, {4 5 8}, {1 5 9}, {1 3 6}, {2 3 4}, {5 6 7}, {3 4 5}, {3 5 6}, {3 5 7}, {6 8 9}, {3 6 7}, {3 6 8}, one need to insert them into a hash tree at first. To generate a candidate hash tree, one need.
- A hash function.
- Max leaf size: max number of itemsets stored in a leaf node (if the number of candidate itemsets exceeds the max-leaf size, split the node).

In this example, we assume the hash function is $hash(X[k]) = X[k] \, mod \, 3$, $X[k]$ is the $k^{th}$ item in the itemset $X$. The hash function is associated with each internal node. Items 1, 4, and 7 are hashed to the same branch (i.e., the middle branch) because they have the same remainder of 1 after dividing the number by 3. The hash tree structure is shown in FIGURE 4-2.

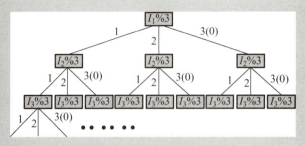

FIGURE 4-2   The hash tree structure

The max-leaf size equals 3. If the number of the leaf node is less than 3, then insert the itemset $X$ to this leaf node. While the number is greater than 3, the leaf node should be split and then store the itemset into the next level of the hash tree. For instance, if we want to insert {1 4 5} into a tree, we should begin with the 1$^{st}$ item {1}. Because 1 *mod* 3 is 1, the itemset {1 4 5} is hashed to the middle child of the root node. Then we can insert the itemset {1 2 4}, {4 5 7}, and {1 2 5} into the same branch because all of the first items in these itemsets have the same remainder of 1. However, when itemset {1 2 5} is inserted in the middle leaf node at level 1, the max-number of the candidate itemsets stored in the child node exceeds 3, so we should split this node and hash them with the 2nd items in these itemsets, i. e. , {4}, {2}, {5}, {2}. Then we store {1 4 5} in the leftmost child node of the left root node at level 1 for the remainder of 4 *mod* 3 is 1, and the rest three itemsets are stored in the middle child node of the same root node at level 1 for the remainder of 2. Next it turns to itemset {4 5 8}, because the first item is {4} and the second item is {5}, it follows the path 1-2 and is hashed into the same leaf node with {1 2 4}, {4 5 7} and {1 2 5}. Because this will violate the limit of maximum leaf size of three, a new splitting operation is applied and the hash tree becomes. Insertion of candiate itemset in FIGURER 4-3.

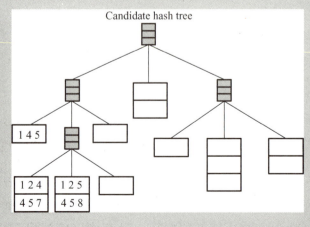

FIGURE 4-3   Insertion of candidate itemsets

In this way, we hash all the candidate itemsets and generate a candidate hash tree of the given dataset shown as in FIGURE 4-4. It contains 15 candidate 3-itemsets, distributed across 9 leaf nodes.

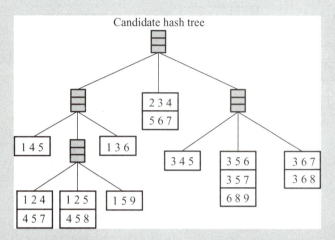

**FIGURE 4-4　A 3-itemset candidates hash tree**

3-itemsets are stored in a dictionary (hash table) with candidate itemsets as keys and the number of occurrence as the count number. With this hash tree, now we can effectively update the support counts of the 3-itemset candidates that belong to a given transaction in the database. Suppose we have a transaction {1 2 3 5 6}, the support counting procedure is different from that of enumerating approach which instead compares each itemset in the transaction with each candidate itemset. During support counting, 3-itemsets contained in each transaction are hashed into their appropriate buckets. At level 1 of the hash tree (FIGURE 4-5), the items {1}, {2}, and {3} are hashed to the left, middle and right child of the root node, respectively. At level 2, the transaction is hashed according to the second item. For example, items {2} and {5} are hashed to the middle child node of the left root node at level 1 (2 $mod$ 3=2, 5 $mod$ 3=2), while item {3} is hashed to the right child node of the same root node as {2} and {5} (3 $mod$ 3=0). This process continues until the leaf nodes of the hash tree are visited. Then the candidate itemsets stored at the visited leaf nodes are matched against the transaction. If a candidate belongs to a subset of 3-itemset of the transaction, its support count is incremented. In this example, a total of 5/9 leaf nodes are visited and 9/15 itemsets are matched against the candidate in the hash tree and the support counts of {1 3 6}, {1 2 5}, and {3 5 6} are updated.

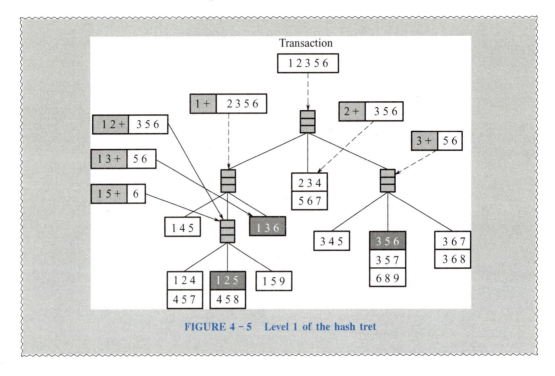

FIGURE 4-5　Level 1 of the hash tret

### 4.3.2　Rule generation

Next, we turn to the introduction of how to generate association rules using the large itemsets which are discovered in the first phase. Every large itemset, $L$, can be partitioned into two non-null subsets, $X$ and $Y(=L-X)$ that describe the subset of $L$ without $X$, thus $X \cap Y = \varnothing$. For association rules, $X \Rightarrow Y$, the key metrics are rule support and confidence predefined by the user. Recall the definition of confidence:

$$Confidence(X \Rightarrow Y) = \frac{\sigma(X \cup Y)}{\sigma(X)} = \frac{Support(X \cup Y)}{Support(X)}$$

a strong association rule must satisfy both *minsup* and *minconf* values. Indeed, all rules have already fulfilled the *minsup* threshold requirement because they are generated from large itemsets.

Another question is that how many rules can be generated from one large itemset? For each large itemset, we can generate so many rules. If a rule $X \Rightarrow Y$ has a confidence value less than *minconf*, then any rule $X' \subset X \Rightarrow (L-X')$ must have a lower confidence value that is greater than the confidence threshold as well. Take the task of mining strong rules from TABLE 4-1 for example. The rule {Bread, PeanutButter}=>{Jelly} has lower confidence, any other rules consisting of the same items and with {Jerry} on the consequent side would have lower confidence too. Specifically, both the rules {Bread => Jelly, PeanutButter} and {PeanutButter => Jelly, Bread} have low confidence actually. Based on this property, all the rules containing an item with low confidence in the consequent can be pruned, so the number of candidate rules that need to be tested will be reduced significantly.

The Apriori algorithm employs a method called level-wise for generating association rules. Initially, we start to generate a list of sets with one item on the right-hand side and test them, and then we merge the left rules meeting the *minconf* requirement to create a new list of candidate rules with two items on the consequent. The process continues until we extract a list of rules with one item on the left-hand side. We illustrate the process of generating rules based on a sample dataset in EXAMPLE 4-3.

**EXAMPLE 4-3**

Suppose we have a large 3-itemset $L_3 = \{a, b, c\}(4)$ with a support count of 4 given in the parathesis, then there are five nonempty subsets which are $\{a, b\}(8)$, $\{a, c\}(4)$, $\{b, c\}(4)$ and $\{a\}(10)$, $\{b\}(9)$, $\{c\}(6)$. Therefore, by applying the level-wise approach, we can get all six candidate rules as follows:
(1) $\{a, b\} => \{c\}$, confidence $= 4/8 = 50\%$.
(2) $\{a, c\} => \{b\}$, confidence $= 4/4 = 100\%$.
(3) $\{b, c\} => \{a\}$, confidence $= 4/4 = 100\%$.
(4) $\{a\} => \{b, c\}$, confidence $= 4/10 = 25\%$.
(5) $\{b\} => \{a, c\}$, confidence $= 4/9 = 44\%$.
(6) $\{c\} => \{a, b\}$, confidence $= 4/6 = 67\%$.

Let the *minconf* $= 70\%$, we only select the second and third rules as the final output for their confidence value are greater than predefined 70%. Note usually for simplicity, single consequent rules are expected.

Lastly, it should be pointed out that not all the rules discovered are meaningful and useful, which would be misleading for business management, especially when there is no sufficient information to determine the relationship of causality implied by the rule.

## 4.4 Frequent Pattern Growth Algorithm

As is introduced above that the Apriori algorithm focuses on discovering large itemsets, and can reduce the number of itemsets to be checked in a database by using the large itemset property. However, Apriori has two main shortcomings that are:
(1) At each step, candidate itemsets have to be created, these itemsets may be large in number if the database is huge.
(2) The algorithm has to scan the database multiple times to examine the support of each candidate itemset, which leads to an expensive cost of computation and needs high memory capacity, especially when the transaction dataset is huge or when the support threshold is low.

Frequent pattern (FP) growth algorithm was proposed to overcome these shortfalls of the Apriori algorithm. A large itemset is generated without considering the candidate generation that is essential for Apriori. FP growth algorithm scans a transaction database only twice and transforms the transaction databases into a tree called FP-tree that compresses a large database into a compact structure. With this approach, the mining of large itemsets is reduced comparatively, which outperforms the Apriori by several orders of magnitude. So it is an efficient and scalable approach for mining full sets of rules in large databases.

To understand how the FP-Growth algorithm helps in finding large items, we first have to know the data structure used by the FP-tree. The root node represents *null* while the leaf nodes represent the itemsets. The tree is constructed by reading the dataset transaction by transaction and mapping each of them onto a path in the FP-tree structure. Note different transactions may have the same paths overlapped for that. They sometimes have items in common. The FP-tree construction is demonstrated in EXAMPLE 4-4.

 EXAMPLE 4-4

Consider the following sample data below.

| Transaction ID | Item |
| --- | --- |
| 1 | {E K M N O Y} |
| 2 | {D E K N O Y} |
| 3 | {A E K M} |
| 4 | {C K M U Y} |
| 5 | {C E I K O O} |

To build the FP-tree, we follow the steps.

**Step 1:** Compute the frequency of individual items below.

| Item | Frequency | Item | Frequency |
| --- | --- | --- | --- |
| A | 1 | M | 3 |
| C | 2 | N | 2 |
| D | 1 | O | 3 |
| E | 4 | U | 1 |
| I | 1 | Y | 3 |
| K | 5 | | |

Let the *minsup*=3. A large 1-itemset $L$ is created containing all the items whose occurrence is greater than or equal to 3. Note that all elements are stored in descending order of their respective support count. Then $L$ looks like, $L=\{K: 5, E: 4, M: 3, O: 3, Y: 3\}$.

**Step 2:** Now, sort items in each transaction in support count descending order, and items with support count value less than 3 are removed from transactions. The following table is built for all the transactions.

| Transaction ID | Item | Ordered itemset |
|---|---|---|
| 1 | {E K M N O Y} | {K E M O Y} |
| 2 | {D E K N O Y} | {K E O Y} |
| 3 | {A E K M} | {K E M} |
| 4 | {C K M U Y} | {K M Y} |
| 5 | {C E I K O O} | {K E O} |

**Step 3:** Create the root node of an FP-tree labeled by "*null*". We will iterate all transactions one by one and insert them into the tree structure.

(1) Insert the first sorted itemset {K E M O Y}. All the items are simply linked after another and the support count for each item as 1. The below figure shows the FP-Tree created after the first transaction has been fully inserted.

(2) Insert the second set {K E O Y}. After the insertion of the K and E, their support counts are simply increased by 1. We will create a new branch for O for that. There is no direct link between E and O. We further insert a new node for the item Y with support count as 1 and link the new node of O with the new node of Y. Then we will have the FP-tree as shown in the figure below.

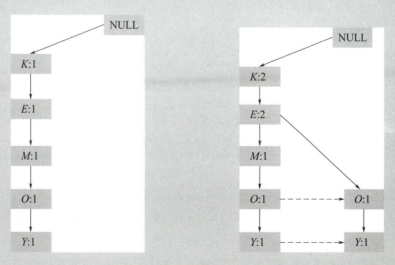

(3) Insert the third set {K E M}. In this step, the support count of each item is increased by 1 and we get the FP-tree updated.

(4) Insert the set {K M Y}. Similar to step (3), first the support count of K is increased by 1, then new nodes for M and Y are created and linked accordingly. the follow figures.

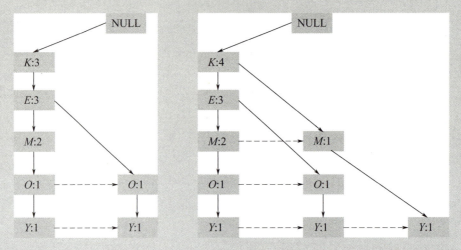

(5) Insert the set of {K E O}. In this step, no new node is created, and we just increase the support counts of the respective items. The FP-tree finally looks like the follow figure.

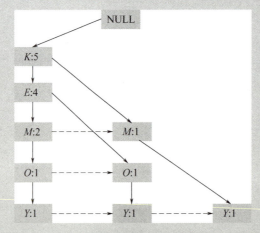

Note in the FP-tree, it contains a list of pointers connecting between nodes that have the same items represented as dashed lines in the above figures, which can be used to facilitate the rapid access of individual items in the tree.

**Step 4**: To extract largeitemsets from the FP-tree. FP-growth algorithm generates frequent itemsets from an FP-tree by examining the tree in a bottom-up manner, that is the algorithm will discover all large itemsets in reverse order. In this example, we first find a large itemset ending of $Y$, followed by $O$, $M$, $E$, and $K$ lastly. Now for every item, we compute their conditional pattern base which is path labels leading to the item in the FP-tree (prefix path ending with each item). Now for each item, the conditional FP-tree is constructed. It is finished by taking the set of elements, which is common in all the paths in the conditional pattern base of that item and calculates its support count by summing the support counts of all the paths in the conditional pattern base.

| Item | Conditional Pattern Base | Conditional FP-tree |
|---|---|---|
| Y | {K E M O: 1}, {K E O: 1} {K M: 1} | {K: 3} |
| O | {K E M: 1} {K E: 2} | {K E: 3} |
| M | {K E: 2} {K: 1} | {K: 3} |
| E | {K: 4} | {K: 4} |
| K | ∅ | ∅ |

Lastly, large itemsets of different lengths are generated by pairing the items in the column of the Conditional FP-tree to the corresponding items in the first column of the table below.

| Item | Largeitemset |
|---|---|
| Y | {K Y: 3} |
| O | {K O: 3} {E O: 3} {K E O: 3} |
| M | {K M: 3} |
| E | {K E: 4} |
| K | ∅ |

For each row, two types of association rules can begenerated. For example, for the first row, two rules K=>Y and Y=>K can be created. To determine a strong rule, the confidence of both the rules is calculated, in this case, they are 60% (3/5) and 100% (3/3), and the rule Y=>K is retained when the *minconf* is 70%.

## 4.5 Measuring the Quality of Association Rules

Millions of rules can be generated by association analysis algorithms. However, some of them might be uninteresting and spurious. Weeding out uninteresting association rules is very important for the successful application of association rule mining. Although Support and confidence are the normal methods used to measure the quality of an association rule, many of the retained rules are still not interesting to users, which is particularly true when mining at low minsup or mining for long patterns. There are some weaknesses associated with these metrics. A strong rule may be obvious but not necessarily interesting. For example, if someone purchases butter, there may be a high likelihood for him/her to Bread. The rule {Buttrer}=>{Butter} is not really of interest because it is not surprising. The limitations of confidence can be traced to the fact that the measure ignores the support of the itemset in the consequent of the rule.

To deal with these issues, a correlation measure can be used to enlarge the framework for association rules. Lift is a simple correlation metric that can be expressed as follows:

$$lift(A \Rightarrow B) = \frac{P(A,B)}{P(A)P(B)} = \frac{confidence(A \Rightarrow B)}{support(B)} = \frac{support(A \cup B)}{support(A) \times support(B)}$$

This measure takes into account both $P(A)$ and $P(B)$. If $lift$ is less than 1, then the occurrence of $A$ is negatively correlated with the occurrence of $B$. In other words, the purchase of $A$ would decrease the probability of purchasing $B$. If the value of $lift$ is greater than 1, then $A$ and $B$ are positively correlated, indicating the occurrence of $A$ implies occurrence of the other. If the value equals 1, then $A$ and $B$ are independent, meaning that there is no correlation between them. Note that one problem with $lift$ measure is that there is no difference between the value for the rule $A \Rightarrow B$ and the value for the rule $B \Rightarrow A$.

Another measure to evaluate the significance of rules is to use the $\chi^2$ test that has been applied in the statistics field. Unlike the support or confidence measurement, this metric takes into account both the presence and the absence of items in itemsets, used to measure how much an itemset (potential correlation rule) count differs from the expected. Given any possible itemset $X$, and an itemset $I = \{I_1, I_2, \ldots, I_m\}$, the $\chi^2$ statistic can be calculated as:

$$\chi^2 = \sum_{X \in I} \frac{(Observed[X] - Expected[X])^2}{Expected[X]}$$

where $Observed[X]$ is the count of the number of transactions that contain the items in $X$; the expected value $[X]$ is calculated as

$$E[X] = n \times \prod_{i=1}^{m} \frac{E[I_i]}{n}$$

Here $n$ is the number of transactions.

## 4.6 Mining Large Itemsets and Association Rules with R Package

In this section, the association rule mining algorithm is implemented with R packages called arules and arulesViz (for visualization). The arules package provides the infrastructure for representing, manipulating, and analyzing transaction data and patterns using large itemsets and association rules. The apriori command in the *arules* mines large itemsets, association rules and class association rules using the Apriori algorithm. Our objective is to review the algorithm theory and demonstrate how to implement it in the *arules* package. We will be following an association rule mining procedure of the "Groceries" database as follows:

(1) Perform exploratory data analysis.

(2) Read thesample transaction dataset in *arules*.

(3) Generate large itemsets in *arules*.

(4) Rule generation in *arules*.

(5) Measuring quality of rules.

- Data exploring

First of all, we should load the required packages *arules* and *aruelsviz* into R Studio.

```
library ("arules")        ## Loading required package
library ("arulesViz")     # visualizing association rules
data ("Groceries")        # To read the transcation dataset from arules package
```

You can use the command inspect () to view the transactions, such as:

```
inspect (head(Groceries)) ## you get the following output
    items
[1] {citrus fruit,
     semi-finished bread,
     margarine,
     ready soups}
[2] {tropical fruit,
     yogurt,
     coffee}
[3] {whole milk}
[4] {pip fruit,
     yogurt,
     cream cheese ,
     meat spreads}
[5] {other vegetables,
     whole milk,
     condensed milk,
     long life bakery product}
[6] {whole milk,
     butter,
     yogurt,
     rice,
     abrasive cleaner}
```

Run the command, *Groceries* and the R help document displays the following information.

The Groceries dataset contains 1 month (30 days) of real-world point-of-sale transaction data from a typical local grocery outlet. The data set contains 9835 transactions and the items are aggregated to 169 categories.

Moreover, you can use the *summary* () to show basic statistics of the dataset. The dataset is rather sparse with a density of just above 2.6%. "whole milk" is the most popular item and the average transaction contains less than 5 items.

```
summary(Groceries) # You may get the following ouput
  transactions as itemMatrix in sparse format with
  9835 rows (elements/itemsets/transactions) and
  169 columns (items) and a density of 0.02609146
most frequent items:
  whole milk other vegetables   rolls/buns      soda      yogurt
       2513            1903          1809       1715        1372
       (Other)
        34055
```

- Read sample dataset

Note that if you need to read data from file as transactions data, use *read.transactions* ().

```
transactdata <- read.transactions("basket.txt", sep="\t") #.
```

After reading the "*transactdata*" into RStudio, we can have a look at the dataset by using str() function.

```
str(transactdata)# #
Formal class 'transactions' [package "arules"] with 3 slots
  ..@ data: Formal class 'ngCMatrix' [package "Matrix"] with 5 slots
  .. .. ..@ i: int [1: 5] 4 0 3 2 1
  .. .. ..@ p: int [1: 6] 0 1 2 3 4 5
  .. .. ..@ Dim: int [1: 2] 5 5
  .. .. ..@ Dimnames: List of 2
  .. .. .. ..$ : NULL
  .. .. .. ..$ : NULL
  .. .. ..@ factors: list ()
  ..@ itemInfo  :'data.frame':   5 obs. of  1 variable:
  .. ..$ labels: chr [1: 5] "a, b, c, f, l, m, o" "a, f, c, e, l, p, m, n" "b, c, k, s, p" "b, f, h, j, o"…
  ..@ itemsetInfo:'data.frame':   0 obs. of  0 variables
```

- Large itemset generation

The *eclat*() takes in a transaction object and gives the most frequent items in the dataset based on the support you provided to the support argument. The *maxlen* defines the maximum number of items in each itemset of large items.

```
frequentItems <- eclat (Groceries, parameter = list(supp = 0.07, maxlen = 15))
# calculates support for large items
inspect(head(sort(frequentItems, by= "support", decreasing= TRUE), 5)) # The
first 5 outputs in decending order are shown as:
   items                  support   transIdenticalToItemsets count
[1] {whole milk}         0.2555160 2513                      2513
[2] {other vegetables}   0.1934926 1903                      1903
[3] {rolls/buns}         0.1839349 1809                      1809
[4] {soda}               0.1743772 1715                      1715
[5] {yogurt}             0.1395018 1372                      1372
```

Then we can plot these frequent items by using the function of *itemFrequencyPlot* (). Item frequency of all items shown in FIGURE 4 – 6.

```
itemFrequencyPlot(Groceries, topN= 10, type= "absolute", main= "Item Occur-
rence")
```

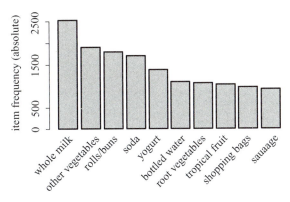

**FIGURE 4 – 6    Item frequency of all items**

- Rule generation

To find the strong association rules, we need to pass values to the parameters of support, confidence. The following script will return to rules whose *minsup* is 0.001 and *minconf* is 0.5.

```
rules <- apriori (Groceries, parameter = list(supp = 0.001, conf = 0.5))
# show the support, confidence and lift for all rules
inspect(head(sort(rules, by= "support", decreasing= TRUE), 5))

    lhs                                              rhs              support
[1] {rice, sugar}                                 => {whole milk}     0.001220132
[2] {canned fish, hygiene articles}               => {whole milk}     0.001118454
[3] {root vegetables, butter, rice}               => {whole milk}     0.001016777
[4] {root vegetables, whipped/sour cream, flour}  => {whole milk}     0.001728521
```

```
    [5]  {butter, soft cheese, domestic eggs}     => {whole milk}  0.001016777
inspect (head (sort (rules, by= "confidence", decreasing= TRUE), 5))
     lhs                                      rhs              support         confidence
[1] {rice, sugar}                          => {whole milk} 0.001220132         1
[2] {canned fish, hygiene articles}        => {whole milk} 0.001118454 1
[3] {root vegetables, butter, rice}        => {whole milk} 0.001016777 1
[4] {root vegetables, whipped/sour cream, flour} => {whole milk} 0.001728521 1
[5] {butter, soft cheese, domestic eggs}   => {whole milk} 0.001016777 1
inspect (head (sort (rules, by= "lift", decreasing= TRUE), 5))
     lhs                                     rhs                    lift
[1] {Instant food products, soda}          => {hamburger meat}      18.99565
[2] {soda, popcorn}                        => {salty snack}         16.69779
[3] {flour, baking powder}                 => {sugar}               16.40807
[4] {ham, processed cheese}                => {white bread}         15.04549
[5] {whole milk, Instant food products}    => {hamburger meat}      15.03823
rules
set of 5668 rules
```

There is a set of 5668 association rules. However, it is obvious that going through all the 5668 rules manually is not an advisable choice. Sometimes it is desirable to remove the rules that are subsets of larger rules. To do so, use the below codes to filter the redundant rules.

```
subsetRules <- which(colSums(is.subset(rules, rules)) > 1) # get subset rules in vector
length(subsetRules)  # > 3913
rules <- rules[-subsetRules] # remove subset rules.
```

- Measuring the quality of rules

```
summary(rules)
set of 5668 rules

rule length distribution (lhs + rhs):sizes
  2    3    4    5    6
 11 1461 3211  939   46

  Min. 1st Qu. Median   Mean 3rd Qu.   Max.
  2.00    3.00   4.00   3.92    4.00   6.00


summary of quality measures:
   support            confidence          coverage            lift              count
 Min.   :0.001017   Min.   :0.5000    Min.   :0.001017   Min.   : 1.957    Min.   : 10.0
 1st Qu.:0.001118   1st Qu.:0.5455    1st Qu.:0.001729   1st Qu.: 2.464    1st Qu.: 11.0
 Median :0.001322   Median :0.6000    Median :0.002135   Median : 2.899
```

```
Median: 13.0
Mean:0.001668    Mean:0.6250    Mean:0.002788    Mean: 3.262
Mean: 16.4
3rd Qu.:0.001729  3rd Qu.:0.6842  3rd Qu.:0.002949  3rd Qu.: 3.691
3rd Qu.: 17.0
Max.:0.022267    Max.:1.0000    Max.:0.043416    Max.:18.996
Max.:219.0


mining info:data transactions support confidence
      Groceries        9835      0.001      0.5
```

The summary shows that:

- √ 5668 rules with a length between 2 and 6 items
- √ range of support: 0.01017—0.022267
- √ range of confidence: 0.5—1.0
- √ range of lift:1.957—18.996

```
subrules2 <-head(sort(rules, by= "lift"), 5)
plot (subrules2, method= "graph")
```

Several purchase rules can be observed. For example:

- √ The most popular transaction is of processed chees, ham and bread
- √ If someone buys{*instant food products*, *soda*}, he is likely going to buy {*hamburger meat*} as well

By visualizing these rules and plots see FIGUTRE 4-7, we can propose more useful insights of how to make business decisions in retail environments and boost store sales.

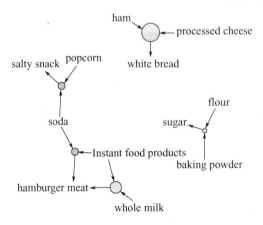

**FIGURE 4-7  Graph-based visualization with rules and plots as vertices**

Note: Larger circles imply higher support, while red circles imply higher lift.

# EXERCISES

1. Given a simple transactional database $X$.

| TID | Item |
|---|---|
| 1 | A B C D |
| 2 | A C D F |
| 3 | C D E G A |
| 4 | A D F B |
| 5 | B C G |
| 6 | D F G |
| 7 | A B G |
| 8 | C D F G |

The threshold values of support and confidence are 25% and 60%, respectively. Please:

(a) find all large itemsets in database $X$.

(b) find strong association rules for database $X$.

(c) analyze misleading associations for the rule set obtained in (b).

2. Find FP-tree for the database X of exercise 1 assuming the support threshold is 3.

3. The following table is called a contingency table, showing the distribution of itemsets with a total sample size of 100 items. Please answer the questions below.

|  | $B$ | $\overline{B}$ | Total |
|---|---|---|---|
| $A$ | 15 | 10 | 25 |
| $\overline{A}$ | 55 | 20 | 75 |
| Total | 70 | 30 | 100 |

(1) $Expected\ [AB]$, $Expected\ [A\overline{B}]$, $Expected\ [\overline{A}B]$, and $Expected\ [\overline{A}\,\overline{B}]$.

(2) Using these above values, please calculate $\chi^2$ for this example and check the significance of the rule $A \Rightarrow B$.

4. Suppose we have the following Table with candidates and large itemsets generated by Apriori algorithm. Please build a hash tree structure to store candidate 3-itemsets. Note each item can be replaced with its numeric value in order, such as Blouse is 1, Jeans is 2, and so on.

| Scan | Candidates | Large Itemsets |
|---|---|---|
| 1 | {Blouse}, {Jeans}, {Shoes}, {Shorts}, {Skirt}, {T-Shirt} | {Jeans}, {Shoes}, {Shorts}, {Skirt}, {T-Shirt} |
| 2 | {Jeans, Shoes}, {Jeans, Shorts}, {Jeans, Skirt}, {Jeans, T-Shirt}, {Shoes, Shorts}, {Shoes, Skirt}, {Shoes, T-Shirt}, {Shorts, Skirt}, {Shorts, T-Shirt}, {Skirt, T-Shirt} | {Jeans, Shoes}, {Jeans, Shorts}, {Jeans, T-Shirt}, {Shoes, Shorts}, {Shoes, T-Shirt}, {Shorts, T-Shirt}, {Skirt, T-Shirt} |
| 3 | {Jeans, Shoes, Shorts}, {Jeans, Shoes, T-Shirt}, {Jeans, Shorts, T-Shirt}, {Jeans, Skirt, T-Shirt}, {Shoes, Shorts, T-Shirt}, {Shoes, Skirt, T-Shirt}, {Shorts, Skirt, T-Shirt} | {Jeans, Shoes, Shorts}, {Jeans, Shoes, T-Shirt}, {Jeans, Shorts, T-Shirt}, {Shoes, Shorts, T-Shirt} |
| 4 | {Jeans, Shoes, Shorts, T-Shirt} | {Jeans, Shoes, Shorts, T-Shirt} |
| 5 | ∅ | ∅ |

5. All the algorithms described in this chapter are mainly for static databases, however, in reality, the transaction databases are undergoing continuous changes. How to mine the association rules is a hard nut to crack. Perform a survey on the web to find recent research techniques for generating rules incrementally.

本章配套资源

# Chapter 5

# Artificial Neural Networks

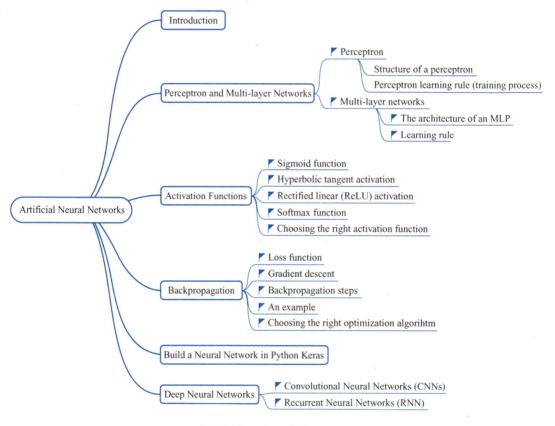

**Mind Mapping of Chapter 5**

人工神经网络（Artificial Neural Network，ANN），也称神经网络（Neural Network，NN），是基于许多被称为神经元的连接处理单元或节点的信息处理系统，可以看作是具有输入（源）、输出（接收器）和隐藏（内部）节点的有向图，且这些节点在不同层之间相互连接。比如，一个简单的3层ANN可以包括一个输入层、一个输出层和一个隐藏层，每层又包括不同数量的神经元。输入层的神经元的树突对应为输入节点，输出层的细胞核处理数据并通过轴突转发计算输出，树突的宽度与每个输入数据的相关权重成正比。

为了完成数据挖掘任务，ANN采用输入元组并训练自己以识别隐藏在数据集中的模式，然后预测新的类似数据集的输出。在此过程中，ANN 的结构不会改变，并根据学习规则（算法）进行更新权重，以及将根据某些条件进行训练和停止。从连通性和学习的角度来看，ANN 通常也被称为前馈和监督学习算法，其中连接仅发生在网络结构中的不同层之间，而在每层内部不连接。

ANN 的主要缺点是它们可能看起来像黑匣子，这意味着很难向最终用户解释中间过程。尽管如此，自 20 世纪 60 年代以来，ANN 的早期成功导致了其可以应用在许多领域，例如计算机视觉、语音识别和自然语言处理等。ANN 可以用于：①函数近似或回归分析，包括时间序列预测和建模；②分类，包括模式识别、新颖性检测和序列决策；③数据处理，包括滤波、聚类、盲信号分离和压缩。另外，需要注意的是，ANN 还提供了深度学习的基础。ANN 可以被视为各种深度学习算法的功能单元，用于解决复杂的数据驱动问题，例如，对数十亿张图像进行分类（比如 Google 图像）、为语音识别服务（比如 Apple 的 Siri）提供支持、每天向数亿用户推荐观看的最佳视频（比如 YouTube）。它们正在被一些知名公司（如 Google、Microsoft 和 Facebook）大规模部署在很多应用场景中。

本章我们将首先简要介绍 ANN 的一些基本概念，包括感知器、多层感知器（Multi-layer Perceptron，MLP）；其次，介绍激活函数和反向传播算法；最后，演示如何使用 Python 中的 Keras 等来构建 ANN 和深度学习架构。此外，还将介绍其他一些深度学习技术。

## 5.1 Introduction

An artificial neural network (ANN), often referred to as a neural network (NN), are developed as a paradigm of modeling the working mechanism of human brains. Such as biological neuron see FIGURE 5-1. ANN is an information processing system based on many connected processing units or nodes called artificial neurons. It can be seen as a directed graph with input (sources), output (sink), and hidden (internal) nodes with in-

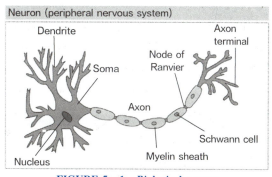

**FIGURE 5-1　Biological neuron**

terconnections between different layers. FIGURE 5-2 is the diagram of a simple 3-layer ANN with one input layer of 4 neurons, one output layer of 3 neurons, and one hidden layer of 6 neurons. By comparing the two figures, we can find that the neurons' dendrites denote as input nodes, and the nucleus process the data and forward the calculated output through the axon. The width of dendrites is proportional to the weight associated with each input data.

To accomplish data mining tasks, ANN take the input tuple and train themselves to recognize patterns hidden in the dataset, and then predict the output for a new similar dataset. During the process, the structure of the graph does not change, the edges (weight) will be updated based

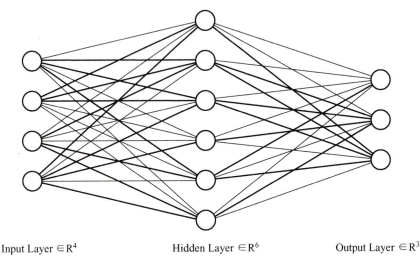

Input Layer $\in R^4$　　　　Hidden Layer $\in R^6$　　　　Output Layer $\in R^3$

FIGURE 5-2　A simple ANN with one hidden layer

on a learning rule (algorithm). NN will be trained and stopped based upon some criteria. We will illustrate the detailed process in the following sections. From the perspective of connectivity and learning, ANN is usually referred to as feedforward and supervised learning algorithm, where connections are only to layers in the graph structure. However, ANN can also be viewed as feedback and unsupervised, where some links are pointed back to previous layers, ever other nodes in the same layer, the networks known as recurrent networks.

A major shortcoming of ANN is that they may look like black boxes, which means that it is difficult to explain to the end-users. Even though, The early success of ANNs since the 1960s leads to many fields that ANN can be applied, such as computer vision, speech recognition, and natural language processing, etc. The tasks to which ANNs are applied include the following categories.

● Function approximation, or regression analysis, including time series prediction and modeling.

● Classification, including pattern and sequence recognition, novelty detection and sequential decision making.

● Data processing, including filtering, clustering, blind signal separation and compression.

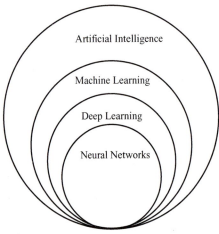

FIGURE 5-3　The relationship between AI, ML, DL and NN

Another important thing to note is that ANNs also provide the basis of deep learning, which is a subfield of machine learning (FIGURE 5-3). An ANN can be regarded as the functional unit of various deep learning algorithms to solve complex data-driven problems, such as classifying billions of images (e.g., Google Images), powering speech recognition services (e.g., Apple's Siri), recommending the best videos to watch for hundreds of millions of users every day (e. g. , YouTube), or learning to beat the world champion at the game of Go by examining millions of past games and then playing against itself (DeepMind's Alpha-Go). So they are being deployed on a large scale by many well-known companies such as Google, Microsoft, and Facebook.

In the following sections of this chapter, we start with a brief tour of some basic concepts of ANNs including perceptrons, MLP, and then introduce activation function, and backpropagation algorithms. Lastly, we will demonstrate how to use deep learning frameworks like Keras in Python to build ANNs and deep learning architectures. Furthermore, some other deep learning techniques will also be introduced. After finishing the study of this chapter, you should be expected to have the ability to apply ANNs and deep learning techniques to solve complex problems.

## 5.2 Perceptron and Multi-layer Networks

### 5.2.1 Perceptron

The perceptron can be thought of as the foundation of ANNs, which was invented by Frank Rosenblatt in 1958 to mimic a biological neuron. To dive into ANNs, it is a good start to have a look at the perceptron algorithm. In the context of ANNs, a perceptron can be viewed as a single-layer feedforward NN which uses the heaviside step function (or unit step function) as the activation function whose value is 0 for negative arguments and 1 for non-negative arguments, where the arguments can be a function that may map its inputs to a binary value. Therefore, a perceptron is actually a binary classifier and can only be used to implement linearly separable classifications. Note that the input layer is not always considered as a real layer of an ANN because the input layer neurons are only to pass and distribute the given data and have no computation performance.

- Structure of a perceptron

A perceptron takes all inputs represented by a vector of numbers and each input node is associated with a weight, then sum the multiplication of each input data with the corresponding weights (called combination function). Afterward, the summed results are fed into an activation function which determines the value (0 or 1) of the output nodes. FIGURE 5-4 is a pictorial representation of the working process of a perceptron. The activation function used in the originally proposed perceptron is the *Heaviside* step function (or unit step function, $f(x)=1$ if $x>0$; 0 otherwise, see EXAMPLE 5-1), but some other types of function are also used instead, such as a *sign* function. A perceptron can compute a continuous output by using a *sigmoid* function: $f(x)=1/(1+e^{-x})$, instead of a step function, which has a continuous derivative allowing it to be used in *backpropagation* (cf. section 5.4). As a matter of fact, one of the main differences between a perceptron and ANN is the activation function. EXAMPLE 5-1 is given to illustrate the process.

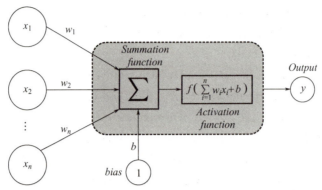

FIGURE 5-4　A pictorial representation of the working process of a perceptron

 **EXAMPLE 5-1**

> Consider a perceptron with two inputs and a bias input. It transforms the inputs $x = (x_1, x_2) = (2, 4)$ into a single output $y$. The weights used are 2, 1, and $-4$ (bias), respectively. Then the weighted value of this perceptron for the input vector is:
> $$S = 2 \times 2 + 4 \times 1 + 1 \times (-4) = 4$$
> By using the unit step function, we get
> $$y = f(S) = \begin{cases} 1 & \text{if } S \geqslant 0 \\ 0 & \text{otherwise} \end{cases}$$
> Here, in this example, we obtain the output value of $y$ is 1 for $S = 4 > 0$. Then the sample belongs to class "1".

● Perceptron learning rule (training process)

Different weights and activation functions can impact the output of a perceptron, and then influence us to make decisions. The goal of a perceptron is to improve the accuracy of classification, which greatly depends on the corresponding weights determined by learning rules. Learning rules are devised to tune the weight values to better approximate some functions that are used for classification or prediction. Therefore, an efficient learning rule of adjusting weights values is to minimize the difference between the projected (predicted) output value ($y^*$) with the targeted (true) output value ($y$), mathematically, it is: $c(w) = |y - y^*|$, which is an error function depending on the weights. The initial learning rule used in a perceptron can be described by the following steps:

**Step 1:** Initialize the weights to 0 or small random numbers.

**Step 2:** For each training input $D = \{(x_1, y_1), (x_2, y_2), \ldots, (x_m, y_m)\}$, $x_i$ is the $n$-dimensional input vector, $y_i$ is the corresponding binary value of output. We can calculate the predicted output value $y_i^*$ by using the perceptron network, then the weight is updated with the format as $w_i := w_i + \Delta w_i$, where the change of weights: $\Delta w_i = \eta(y_i - y_i^*)x_i$, $\eta$ is the learning rate (a constant with a range from 0 to 1) predefined by users and larger values make the weight changes more volatile. A rule of thumb is to set $\eta$ equals the reciprocal of the number of entries in the training input set. It should be pointed out here that all weights in the weight vector are being updated simultaneously, in other words, all these weights are immediately updated while a pair in the training dataset is considered instead of waiting after all pairs are traversed. For instance, for a training dataset with 2-dimensional inputs, the weight is updated as:

$$\Delta w_0 = \eta(y_i - y_i^*)x_{i,0}$$
$$\Delta w_1 = \eta(y_i - y_i^*)x_{i,1}$$
$$\Delta w_j = \eta(y_i - y_i^*)x_{i,2}$$

where, $x_{i,j} = 0, 1, 2$ is the value of the $i$th element of the $j$th training input vector $x_i$.

**Step 3**: Repeat Step 2 until the error $c(w)$ is less than a small number specified by the user or a predetermined number of iterations has been accomplished.

The learning rule (algorithm) is guaranteed to converge to a stable state when the training dataset is linearly separable. Note that although the weights of single-layer perceptrons are easier to train and learn, the weights updating process above is not applicable in the case when the problem modeled by a perceptron is nonlinear because the step activation function is non-differentiable. In the real world, most problems are nonlinear and multiclass. If we want to discover these patterns, alternative models should be developed such as multi-layer networks which will be discussed in the next section.

### 5.2.2 Multi-layer networks

Multi-layer Perceptrons (MLP), sometimes referred to as feedforward ANNs, usually consists of two or more layers of nonlinearly-activating nodes, which have a greater processing power to engage in more sophisticated decision-making problems than a single perceptron. An MLP may also be viewed as a network of multiple perceptrons, so to speak, if there are two or more layers in a neural network it is a multi-layer perceptron, otherwise, it should be called a single perceptron.

FIGURE 5-2 shows a multi-layer perceptron with a single hidden layer. The output of the node in hidden layers depends on the outputs from the input layer and the weights associated with the connections between them. Then these outputs enter into the corresponding connected nodes in the output layer. Note that even the outputs of some neurons can become inputs to other neurons, cycles are not allowed for an MLP.

• The architecture of an MLP

The most common type of MLP architecture is the fully-connected feedforward NN where neurons in two neighboring layers are fully pairwise connected with corresponding weights $w_{ij}$, but neurons within a single layer share no connections.

Another two issues associated with the MLP network structure are the number of hidden layers and the number of neurons in each hidden layer which are model hyperparameters specified by users. The first issue determines the number of parameters (weights) to learn. An MLP with multiple hidden layers may be called a deeper NN as represented in FIGURE 5-5. According to Kolmogorov's theorem, an MLP theoretically needs no more than two hidden layers to perform a mapping between two sets of numbers. MLP with at least one hidden layer are universal approximators, that is to say, arbitrary decision regions can be arbitrarily well approximated by continuous feedforward neural networks with only a single internal, hidden layer and any continuous sigmoidal nonlinearity. In fact, for many practical problems, it is often the case that 3-layer neural networks will have a better performance over 2-layer NN, but going even deeper (4, 5, 6-layer) is rarely helpful.

As far as the number of neurons in eachhidden layer is concerned, it is very important to determine the overall NN architecture. One of the most common rules of thumb is to choose a number of hidden neurons between 1 and the number of input attributes. Larger NN with more neurons can represent more complicated functions. However, overfitting would occur when an MLP with too many neurons in each layer, which leads to a high capacity of fitting the noise in the training dataset other than the (assumed) underlying relationship. This is an annoying issue to approach in many practical problems.

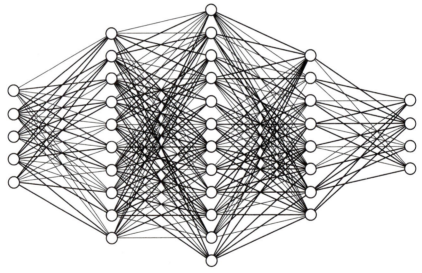

FIGURE 5-5  A deeper neural network

● Learning rule

As to the learning of arguments in MLP, the backpropagation algorithm is one of the most popular algorithms for multilayer networks that makes a significant step forward from the single-layer perceptron network. The learning process starts from the output layer backward to the input layer. In short, *backpropagation* can learn from errors in the output compared to the desired result, and adjust the weights to minimize the total error over the output nodes. Suppose that $f(W, X)$ that represents an MLP, $W$ is the vector of learnable weights, and $X$ is the given input tuples, the *backpropagation* algorithm simply calculates the gradients $\partial f/\partial W$. A simplified *backpropagation* algorithm with the mean squared error (MSE) as the loss function is shown below. For a more detailed introduction, you can directly skip to section 5.4.

A simplified background progagation algorithm.

**Input:**
    $W$ //The weights required to be learned, initially random assigned
    $X = (x_1, x_2, \ldots, x_n)$ //The input vector with $n$ features
    $O = (o_1, o_2, \ldots, o_m)$ //The output vector, $m$ is the number of samples in the training set.

**Output:**
    $W'$ //The updated weights in the trained MLP

1. Propagation $(W, X, O)$.
2. $\boldsymbol{L} = 1/2 \sum_{1}^{m} (o_i - y_i)^2$ ; //$\boldsymbol{L}$ is the loss function. $y_i$ is the predicted output.
3. Gradient $(W, L)$; //The *gradient descent* method to modify the weights.

The basic idea of gradient descent is to tune the weights in each layer to minimize the MSE loss function. For each layer $\ell$, we update: $W^{[\ell]} = W^{[\ell]} + \eta \frac{\partial L}{\partial W^{[\ell]}}$, $\eta$ is the learning rate.

## 5.3 Activation Functions

The output of each node in ANNs depends on the definition of the activation function $f_i$. An activation function is used to perform certain operations to the inputs dataset. Usually, these inputs are linearly combined with their weights. Then the activation function squashes the combination to a range between 0 and 1. Activation functions can induce significant impacts on ANN's ability to converge and the converging speed in some cases. Hence, activation functions are a crucial component of ANNs.

There are many different types of activation functions. In one of Hinton's seminal papers on automatic speech recognition, he used a logistic sigmoid activation function. In the architectures of AlexNet, ResNet and BERT, ReLU and GELU are used as the activation function. In this section, we will introduce several activation functions you may encounter in practice, such as sigmoid, tanh, and ReLU, which are all non-linear and each of them has its own applications.

### 5.3.1 Sigmoid function

A sigmoid function has an "S"-shaped curve that is monotonically increasing (FIGURE 5-6). A typical sigmoid function is a logistic function, defined by the following mathematical formula:

$$f(S) = \frac{1}{(1 + e^{-cS})}$$

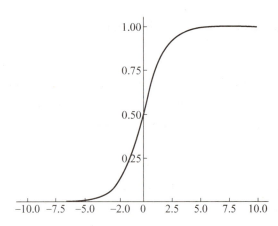

**FIGURE 5-6  Curve of the sigmoid function**

where $c$ is a constant positive value used to change the slope of the function. The output of the function ranges between 0 and 1. When small changes in $S$ would cause large changes in the value of output. Especially, the function can transform large negative numbers into 0 and large positive numbers into 1. This is an attractive property to be applied in binary classifications. The output can be predicted easily to be 1 if the value is greater than 0.5 and 0 otherwise. Moreover, the function possesses a simple derivative: $\frac{\partial f_i}{\partial S} = f_i(1-f_i)$.

However, this function has its own drawbacks. As shown in FIGURE 5-6, the output of the sigmoid function tends to respond less to changes in the variable $S$ for both tails in the horizon axis. The gradient at these regions becomes very small and almost zero, which leads to an issue of the vanishing gradients making the NN hard to learn.

### 5.3.2  Hyperbolic tangent activation

The hyperbolic tangent function (also known as tanh function) is a variation of the sigmoid function. It is defined by:

$$f(S) = \frac{e^S - e^{-S}}{e^S + e^{-S}}$$

This function is represented in FIGURE 5-7. As is shown, the hyperbolic tangent has a similar S shape to the sigmoid function and produces outputs in the scale of $(-1, 1)$. But its output is zero centered. Actually, the *tanh* is a scaled sigmoid function, particularly, $tanh(S) = 2sigmoid(2S) - 1$. The derivative of tanh is $\frac{\partial f_i}{\partial S} = 1 - (f_i)^2$. Note that, like sigmoid function, the *tanh* also has the vanishing gradient problem.

### 5.3.3  Rectified linear (ReLU) activation

The rectified linear activation function, also known as ReLU, is another non-linear activation function that has gained popularity in the machine learning domain with state-of-the-art results. It usually achieves better performance and generalization in deep learning

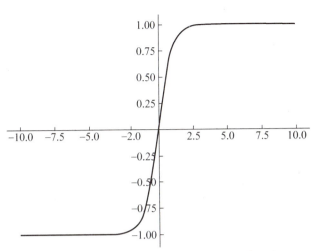

**FIGURE 5-7** Curve of hyperbolic tangent function

compared to the sigmoid activation function. The ReLU function is defined by $f(S)=\max(0,S)$, namely, the activation is simply thresholded at zero. Its curve is shown in FIGURE 5-8, as can be seen, it is half rectified from the bottom, and has an output range of $[0, \inf]$.

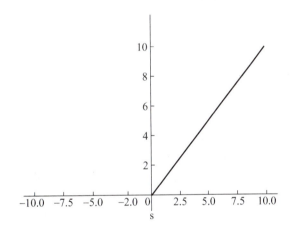

**FIGURE 5-8** Curve of ReLU function

An advantage of the ReLU function is that it is more computationally efficient than other activation functions such as sigmoid and *tanh* due to its linear and nonsaturating form. Furthermore, the ReLU function is involved in fewer operations compared to *tanh* and sigmoid functions. That is a good property to consider when we are designing NNs. In order to approach the "dying ReLU" problem, Leaky ReLU function is proposed with the following improved definition:

$$f(S)=\begin{cases} \alpha S & \text{for } S<0 \\ S & \text{for } S\geqslant 0 \end{cases}$$

where $\alpha$ is a small constant.

### 5.3.4 Softmax function

The softmax function is a generalization of the logistic function to multiple dimensions. It is mostly used as the last activation function of a neural network to normalize the output of a network to a probability distribution over predicted output classes and it is not a function of a single fold of inputs from the previous layer or layers. The function can turn any arbitrary real values into probabilities, which are useful in determining the target class for the given inputs. After applying the softmax function, each output will be limited in the interval (0, 1), the sum of all the outputs probabilities equals 1, then they can be interpreted as probabilities. Moreover, the larger output value will correspond to larger probabilities. The Softmax function is defined using the equation below:

$$P(y=i|f(S_i)) = \frac{e^{f(S_i)}}{\sum_{k=1}^{m} e^{f(S_k)}}$$

The main difference between the sigmoid and softmax functions is that sigmoid is usually applied in binary classification while the softmax can be used for multi-class tasks other than binary classification. A simple example is given to illustrate the function (EXAMPLE 5-2).

### EXAMPLE 5-2

Suppose we have output values $f(S)=[-1, 0, 2, 5]$. Intuitively, the output value of "5" should have the highest probability. We first calculate the denominator of the sigmoid function and get:

$$e^{-1} + e^0 + e^2 + e^5 = 157.1701$$

Then we can calculate the probabilities as follows.

| $f(s_i)$ | $e^{f(S_i)}$ | $P(y=i|f(S_i))$ |
|---|---|---|
| −1 | 0.3678 | 0.00234 |
| 0 | 1.0 | 0.00636 |
| 2 | 7.3891 | 0.04702 |
| 5 | 148.4132 | 0.94428 |
| Sum (rounded) | | 1.00000 |

The greater the $f(s_i)$, the larger its probability, and all the probabilities sum up to 1, shown in the last column.

### 5.3.4 Choosing the right activation function

Another important issue is how to choose the right type of activation function for your NN model. A basic rule of thumb is that if you have no idea of the nature of the pattern of the training dataset and really don't know what activation function should be used, the ReLU is simply a good start as it is a general approximator of most functions. If your output is for binary classification then, *sigmoid* and *tanh* functions are the preferred choices for most of the time. Actually, in many practical applications, the activation functions are selected depending upon the type of problem to be solved by ANNs. Note also it is very rare to apply different types of functions in the same layer, but you can do so in different hidden layers.

Recent research attempted to discover approaches to automatically learn different combinations of base activation functions (such as the sigmoid function, ReLU, and tanh) during the training phase to achieve the highest accuracy. It is a very promising field of research because it attempts to automatically find an optimal configuration of the activation function rather than manually tune it.

## 5.4 Backpropagation

Backpropagation (BP), also known as backward propagation of errors, is a classical learning technique to tune weights in ANNs by propagating weight changes backward from the output/sink to the input/source nodes. Backpropagation can be thought of as a generalized delta rule approach. This section focuses on how to find the optimal weights/bias parameters of an ANN that can minimize the value of a loss function, which is a performance measure evaluated on the training tuples.

### 5.4.1 Loss function

An important aspect of the design of a neural network is the choice of the loss function that is a function that maps values of one or more variables onto a real number intuitively representing some "cost" associated with those values.

For backpropagation, the loss function is used to calculate the difference between the predicted network output and its desired output after a training tuple has propagated through the network. This prediction error is analogous to the residuals in regression models. Note that sometimes loss functions are referred to as cost functions (or error functions) as both of them have to define an objective function to measure the model performance for a given training dataset and the purpose of them is to be either minimized or maximized. Intuitively, the loss will be larger if the model has a poor ability of classification for the training data, otherwise, the loss will be smaller.

There are many different types of loss functions. In this part, we only cover a few commonly used in practice.

- MSE

MSE is one of the most widely used loss functions in many practical applications. The MSE formula can be written as: $L(w,b) = 1/2m \sum_{i=1}^{m}(o_i - y_i)^2$ , where the number 2 in the denominator is just for mathematical convenience to simplify the calculation of the derivative later. To get more precise, the MSE becomes:

$$L(w,b) = 1/2m \sum_{i=1}^{m} (o_i - y_i)^2$$

$$= 1/2m \sum_{i=1}^{m} (o_i - \sigma(WX + b))^2$$

where $o_i$ is the true output value, $y_{i \in m}$ is the predicted output value among $m$ classes, $X = [x_1, x_2, \ldots, x_n]'$ is the input vector, $W$ is the weights vector needed to be optimized, $\sigma()$ is the activation function used which can be substituted by any type described previously.

Note that Mean Absolute Error (MAE) is also used as a very simple loss function. However, to calculate derivatives of MAE is problematic. By contrast, MSE has good mathematical properties which make the computation of its derivative easier. It is especially more convenient when applying the gradient descent algorithm. Especially it is the preferred loss function under the inference framework of maximum likelihood if the distribution of the target variable is Gaussian.

- Cross-entropy loss

Besides the MSE, cross-entropy loss, or *log* loss, is another popular loss function, which stems from the information theory and measures the performance of a model with a probability value [0, 1] as outputs. It is intended for use with binary and multi-class classification problems. Cross-entropy loss has the property that it decreases rapidly as the predicted probability approaches the true label (FIGURE 5-9). A perfect model would have a cross-entropy loss of zero, so predicting a probability of 0.15 when the actual observation label is 1 would be worse than a predicted value of 0.05 and result in a high loss value. The cross-entropy between a "true" probability $o$ and an estimated distribution $y$ is:

$$H(o,y) = -\sum_{i}^{m} o_i \log y_i$$

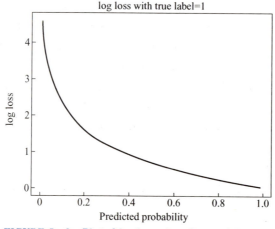

**FIGURE 5-9　Plot of log loss when the true label is 1**

More specifically, for binary classification where the class number equals 2 (usually labeled with 0 and 1), the cross-entropy can be written as:
$$H(o,y) = -o\log y - (1-o)\log(1-y)$$

- Kullback-liebler divergence loss (KL Divergence)

KL Divergence (KLD), also known as relative entropy, is a non-symmetric measure of how a probability one distribution differs from a reference probability distribution. In the simple case, two distributions under consideration are identical if the KLD equals 0. Given $P$ as the observations, or a probability distribution actually measured and $Q$ denoting a representation or an approximation of $P$. $D_{KL}(P \parallel Q)$, reading as divergence from $Q$ to $P$, is interpreted as the information gain when $Q$ is used rather than $P$ and has the definition on a discrete probability space $\chi$:

$$\begin{aligned} D_{KL}(P \parallel Q) &= \sum_{x \in \chi} P(x) \log\left(\frac{P(x)}{Q(x)}\right) \\ &= \sum_{x \in \chi} P(x) \log P(x) - \sum_{x \in \chi} P(x) \log Q(x) \\ &= H(P,Q) - H(P) \end{aligned}$$

where $H(P, Q)$ is the cross-entropy of $P$ and $Q$, and $H(P)$ is the entropy $P$. The goal of the KLD is to approximate the true probability distribution $P$ of the outputs in terms of the inputs with the distribution $Q$. This can be achieved by minimizing the $D_{KL}(P \parallel Q)$ that is called forward KL. The higher the similarity between $P$ and $Q$, the lower the KDL will be. If $P$ and $Q$ perfectly match, $D_{KL}(P \parallel Q) = 0$.

The KLD loss function is more commonly used when using models that learn to approximate a more complex function than simply multi-class classification, such as in the case of an autoencoder used for learning a dense feature representation under a model that must reconstruct the original input.

In summary, we can follow the suggestions below as the issue of how to choose loss functions for a neural network.

(1) Regression Problem: a problem where you predict a real-value quantity.
- Output Layer Configuration: one node with a linear activation unit.
- Loss Function: mean squared error (MSE).

(2) Binary Classification Problem. A problem where you classify an example as belonging to one of two classes.
- Output Layer Configuration: one node with a sigmoid activation unit.
- Loss Function: cross-entropy, also referred to as logarithmic loss.

(3) Multi-Class Classification Problem. A problem where you classify an example as belonging to one of more than two classes.
- Output Layer Configuration: one node for each class using the softmax activation function.
- Loss Function: cross-entropy, also referred to as logarithmic loss.

### 5.4.2 Gradient descent

Gradient descent and its variants are probably the most used optimization algorithms for machine learning and especially for deep learning. Gradient descent involves calculating the derivative of the loss function concerning the weight for each layer in a neural network, so the BP algorithm is also a gradient descent-based learning algorithm/rule. Then understanding gradient descent is critical to effectively comprehend neural networks.

The gradient descent employs the chain rule and product rule in differential calculus which requires the differentiation of the activation and loss functions. Recall that the chain rule is a formula for calculating the derivatives of composite functions which are composed of functions inside other function(s), formally, $f(g(x))' = f'(g(x)) \cdot g'(x)$. The product rule is a formula used to find the derivatives of products of two or more functions (or composite functions), mathematically, $(f(x) \cdot g(x))' = f'(x) \cdot g(x) + f(x) \cdot g'(x)$. The chain structures are the commonly used structures of neural networks. For example, there are three functions $f_1$, $f_2$, $f_3$, connected by three layers in turn, then the final output can be written as $f(WX + b) = f_3(f_2(f_1(X)))$.

Before presenting gradient descent, we should define some notations which are used to refer to some parameters in the neural network.

- Dataset $D$, composed of input-output pairs $(x_i, o_i)$, $i = 1, 2, \ldots, m$, where the input vector $X_i$ has $n$ dimensions, $m$ is the size of samples in the dataset.
- Loss function $L(W, b)$, defining the error between the desired output $o_i$ and predicted output $y_i$ of the neural network.
- The parameters to be optimized, consisting of vectors of $W$ and $b$, where each entry of $W$ is $w_{pq}^k$ denoting that the weight between the $q$th node in the $(k-1)$th layer and the $p$th node in the $k$th layer. $b_p^k$ is the bias for node $p$ in the $k$th layer.
- The output for node $p$ in layer $k$, denoting by $a_p^k = \sigma(s_p^k)$, $s_p^k = \sum_q w_{pq}^k a_q^{k-1} + b_p^k$. For the input layer, $k = 0$, $a_q^0 = x_i^q$, where $x_i^q$ is the $q$th element of the input vector $X_i$. In FIGURE 5-10, we can obtain $a_1^k$ as: $a_1^k = \sigma(a_1^2 \times w_{11}^2 + a_2^2 \times w_{12}^2 + a_3^2 \times w_{13}^2 + b_1^k)$.

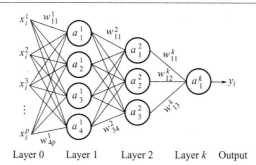

FIGURE 5-10  A simple representation of a $k$-layer neural network

The diagram above shows these notations in a simplified network.

To train ANNs, the core idea of gradient descent is to iteratively update the weights and bias parameters by taking steps proportional to the negative of the gradient of the loss function $L(W, b)$ with respect to $w_{pq}^k$ and $b_p^k$ (collectively denoted with $\theta$): $\theta(t+1) := \theta(t) - \eta \Delta\theta(t)$, where $\eta$ is the learning rate representing the step size moving along the negative direction of the gradient, $\theta(t)$ is the value of weight and bias $(w, b)$ at iteration $t$, $\Delta\theta$ is defined as: $\Delta\theta = [\Delta w, \Delta b] = \left[\frac{\partial L}{\partial w}, \frac{\partial L}{\partial b}\right]$. The goal of applying gradient descent to the loss function is to find an optimal combination of weights/bias to make the difference between the predicted outputs and expected values as small as possible. We illustrate the process of deriving the gradient by using the loss function of MSE, i.e. $L(W, b) = \frac{1}{2m} \sum_{i=1}^{m} (o_i - y_i)^2$, thus the goal is: $\operatorname{argmin}_\theta \frac{1}{2m} \sum_{i=1}^{m} (o_i - y_i)^2$. Notice that the following deduction of formulations used is only for a simple NN with one output, that is, $o_i$ and $y_i$ are real values other than vectors. However, the process can be extended to other NN models with any number of outputs by applying the chain rule and product rule during derivation.

It should also be pointed out that the above loss function is a total error function after all training samples are processed, which can be decomposed into a sum of all individual error terms. Since the derivative of a sum of functions is the sum of the derivatives of each function, so we can start calculating derivatives with one individual sample, and then obtain the general form of all training sets. Hence, by using the chain rule, the derivation of the loss function $L_i$ with respect to $w_{pq}^k$ for sample $i$ is:

$$\frac{\partial L_i}{\partial w_{pq}^k} = \frac{\partial L_i}{\partial y_i} \cdot \frac{\partial y_i}{\partial w_{pq}^k} = \frac{\partial L_i}{\partial y_i} \cdot \frac{\partial y_i}{\partial s_p^k} \cdot \frac{\partial s_p^k}{\partial w_{pq}^k}$$

where the second factor of the right-hand side $\frac{\partial y_i}{\partial s_p^k}$ is the partial derivative of the activation function, namely, $\frac{\partial y_i}{\partial s_p^k} = \frac{\partial a_p^k(s_p^k)}{\partial s_p^k}$, if the activation function is the sigmoid function, then it can be written as: $\frac{\partial a_p^k(s_p^k)}{\partial s_p^k} = y_i(1 - y_i)$. The third factor $\frac{\partial s_p^k}{\partial w_{pq}^k}$ can be calculated as:

$$\frac{\partial s_p^k}{\partial w_{pq}^k} = \frac{\partial \left(\sum_q w_{pq}^k a_q^{k-1} + b_p^k\right)}{\partial w_{pq}^k} = a_q^{k-1}$$

Now for the first factor, it is calculated as: $\frac{\partial L_i}{\partial y_i} = \frac{\partial [1/2(o_i - y_i)^2]}{y_i} = (y_i - o_i)$. Then combining these factors together, we can get the partial derivative of the loss function with respect to a weight in the final layer:

$$\frac{\partial L_i}{\partial w_{pq}^k} = \frac{\partial L_i}{\partial y_i} \cdot \frac{\partial y_i}{\partial w_{pq}^k} = \frac{\partial L_i}{\partial y_i} \cdot \frac{\partial y_i}{\partial s_p^k} \cdot \frac{\partial s_p^k}{\partial w_{pq}^k} = (y_i - o_i) \cdot y_i(1 - y_i) \cdot a_q^{k-1} \quad (5.1)$$

Especially we can replace $k = m$, and $q = 1$ (for there is only one output node for each individual sample), thus the partial derivative can be written as:

$$\frac{\partial L_i}{\partial w_{p1}^m} = \frac{\partial L_i}{\partial y_i} \cdot \frac{\partial y_i}{\partial w_{pq}^m} = \frac{\partial L_i}{\partial y_i} \cdot \frac{\partial y_i}{\partial s_p^m} \cdot \frac{\partial s_p^m}{\partial w_{p1}^m} = (y_i - o_i) \cdot y_i(1 - y_i) \cdot a_1^{m-1}$$

However, when the question involves calculating the derivative of the loss function in an arbitrary hidden layer, it is a little bit complicated. To calculate the partial derivatives, we first define the term $\delta_p^k$ called the error in $p$th node of the $k$th layer, denoted as: $\delta_p^k = \frac{\partial L_i}{\partial s_p^k}$. Then the loss function partial derivative can be written as:

$$\frac{\partial L_i}{\partial w_{pq}^k} = \delta_p^k \frac{\partial s_p^k}{\partial w_{pq}^k} = \delta_p^k a_q^{k-1}$$

This is reasonable since the weight $w_{pq}^k$ connects the output of the node $q$ in the layer $k-1$ to the input node $p$ in the layer $k$ in the BP computation graph. Notice here we do not specify a particular error function or activation function. By applying the chain rule, we can rewrite the term $\delta_p^k$ with respect to $\delta_p^{k+1}$:

$$\delta_p^k = \sum_r \frac{\partial L_i}{\partial s_r^{k+1}} \cdot \frac{\partial s_r^{k+1}}{\partial s_p^k}$$

Recalling that $s_r^{k+1} = \sum_p w_{rp}^{k+1} a_p^k + b_r^{k+1} = \sum_p w_{rp}^{k+1} \sigma(s_p^k) + b_r^{k+1}$, then we differentiate it with respect to $s_p^k$, and obtain:

$$\frac{\partial s_r^{k+1}}{\partial s_p^k} = w_{rp}^{k+1} \sigma'(s_p^k)$$

Then replacing it back into the partial derivative of the loss function, it can be formulated as:

$$\delta_p^k = \sum_r \frac{\partial L_i}{\partial s_r^{k+1}} \cdot \frac{\partial s_r^{k+1}}{\partial s_p^k} = \sum_r \frac{\partial L_i}{\partial s_r^{k+1}} \cdot \frac{\partial s_r^{k+1}}{\partial s_p^k} = \sum_r \frac{\partial L_i}{\partial s_r^{k+1}} w_{rp}^{k+1} \sigma'(s_p^k)$$

Substituting the definition $\delta_r^{k+1}$ for the term $\frac{\partial L_i}{\partial s_r^{k+1}}$ and swapping the positions of them, we get:

$$\delta_p^k = \sum_r \delta_r^{k+1} w_{rp}^{k+1} \sigma'(s_p^k) = \sigma'(s_p^k) \sum_r \delta_r^{k+1} w_{rp}^{k+1}$$

Putting them all together, we can write the partial derivative of the loss function with respect to weights in hidden layers $w_{ij}$ as:

$$\frac{\partial L_i}{\partial w_{pq}^k} = a_q^{k-1} \delta_p^k = a_q^{k-1} \sigma'(s_p^k) \sum_r \delta_r^{k+1} w_{rp}^{k+1} \quad (5.2)$$

Similarly, we can get the equation for the partial derivative of the loss function with respect to the bias in the network as:

$$\frac{\partial L_i}{\partial b_p^k} = \delta_p^k = \sigma'(s_p^k) \sum_r \delta_r^{k+1} + w_{rp}^{k+1} \quad (5.3)$$

Actually, equations 5.2 and 5.3 are the two final formulas for a neural network to update the weights and bias, which are used to decide the moving directions of these important parameters. Therefore, applying the two important equations iteratively can help achieve lower values for the loss function attempting to minimize the gap between the predicted outputs and true outputs. Note that the gradient descent is not guaranteed to find the minimum of the loss function especially when it is not a concave function.

It is important to note that the frequency of the weights updated is influenced by the size of training samples entered into the NN model, hence impacts the accuracy of gradient descent. Usually, there are three modes to train the model from the perspective of sample size. (1) Online or incremental mode. In this mode, the weights are changed after each training sample is entered into the network. This mode is suitable when your training dataset is not large, and absent of outliers. (2) Batch or offline mode. In this mode, the weights are updated after all the training samples are fed to the model, the batch mode of training is more robust to variance in the training dataset. (3) Stochastic or mini-batch mode. In this mode, a number (predefined by the user) of samples are chosen randomly from the training dataset, and the training process is finished in batch mode over these mini-batches. This mode can help avoid getting stuck in local optima.

The definitions of epoch, iteration and batch size are illustrated in the EXAMPLE 5-3.

 **EXAMPLE 5-3**

- One epoch refers to that all training samples are completely passedforward and backward through the network model for one time. In practice, for a massive dataset, one epoch is too large to feed to the computer at once, so we should divide it into several smaller batches.
- Batch size is the number of samples processed before the model parameters are updated. Note that batch size and the number of batches are two different things.
- Iteration means the number of batches needs to complete one epoch. The number of batches is equal to the number of iterations for one epoch.

Assume we have a dataset with 100 samples and set a batch size of 20 and 500 epochs. This means that the dataset will be divided into 5 batches, each with 20 samples. The model parameters (weights and bias) will be updated after each batch of 20 samples. In other words, this also means that for each epoch there will be 5 batches or 5 updates to the model involved. With 500 epochs, the network will process the whole dataset 500 times, i.e., a total of 2500 batches during the entire learning cycle, where the number of iterations is 5 for one entire epoch and total iterations of 2500 times.

### 5.4.3 Backpropagation steps

Generally, the BP algorithm proceeds in the following six typical steps, assuming a suitable learning rate $\eta$, random initialization of the parameters $w_{ij}^k$, number of hidden layers, activation function, loss function, epoch, iterations and batch size.

**Step 1:** Inputs-output pairs $(x_i, o_i), i=1,2,\dots,m$. Randomly choose a batch of samples from the training datasets.

**Step 2:** Forward propagation. The inputs are passed through the neural network and the corresponding output $(y_i)$ is calculated, including $s_p^k$ and $a_p^k$, from the input to the output layer.

**Step 3:** Output error calculation. Calculate partial derivatives from the loss function to a specific neuron and its weight. Specifically, get the error terms for hidden layers and the output layer using equation 5.1 to equation 5.3.

**Step 4:** Combine the gradient for each input-output pair to get the total gradient for a batch of training dataset $\frac{\partial L}{\partial w_{pq}^k} = \frac{1}{mb} \sum_i \frac{\partial L_i}{\partial w_{pq}^k}$ and $\frac{\partial L}{\partial b_p^k} = \frac{1}{mb} \sum_i \frac{\partial L_i}{\partial b_p^k}$, $mb$ is the batch size.

**Step 5:** Weights and bias update. Multiplying the learning rate $\eta$ with the total gradient obtained in Step 4 and moving along the negative direction of the gradient. These parameters are changed to a new value according to the results of the backpropagation algorithm.

**Step 6:** Iterate until convergence. Repeat Step 1 to Step 5 to train the network towards less and less global loss function until the algorithm converges to a small value or meets other stopping criteria such as iteration number or loop number.

After finishing the above steps, the network is trained for making predictions for any new input data fed into the model, then the predictions are generated for decision making.

### 5.4.4 An example

To get an intuitive understanding of BP, we present a concrete example to illustrate the algorithm below. In this example, suppose we have a neural network with one hidden layer and two neurons in the output layer. Its structure is shown in FIGURE 5-11.

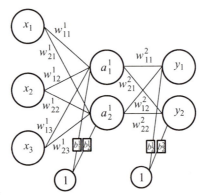

FIGURE 5-11  The structure of the NN in the example

Assume that the training dataset includes one single sample: the inputs vector $x=(x_1, x_2, x_3)$, is set as [1, 4, 5], the outputs $O=(o_1, o_2)$ are expected to be [0.1, 0.05]. As alluded to in the previous section, the goal of backpropagation is to find the optimal weights to minimize the difference between the predicted outputs and expected values. Note that even we use only one hidden layer in this example, the BP process also is applicable when more hidden layers are required to be dealt with.

Firstly, we start by the forward propagate through the network to get the predicted output values $y_i$. To do this we should feed the inputs forward through the network. Following the notation used in the gradient descent, we denote the value before the activation function (i.e., the sigmoid function in this example) is applied with $s_1^1$, $s_2^1$, $s_1^2$, $s_2^2$, and denote the value after the activation function is applied with $a_1^1$, $a_2^1$, $y_1$, $y_2$. Then we calculate the outputs of hidden layers with the initial values in FIGURE 5 - 12, by using the formulas in section 5.4.2, the results are given as.

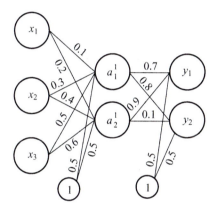

**FIGURE 5 - 12  Initialization of the NN**

for $a_1^1$ and $s_1^1$:
$$s_1^1 = w_{11}^1 x_1 + w_{12}^1 x_2 + w_{13}^1 x_3 + b_1^1 = 0.1 \times (1) + 0.3 \times (4) + 0.5 \times (5) + 0.5 = 4.3$$
$$a_1^1 = \sigma(s_1^1) = \sigma(4.3) = 0.9866$$

Carrying out the same process for $a_2^1$ and $s_2^1$, we get:
$$s_2^1 = w_{21}^1 x_1 + w_{22}^1 x_2 + w_{23}^1 x_3 + b_2^1 = 0.2 \times (1) + 0.4 \times (4) + 0.6 \times (5) + 0.5 = 5.3$$
$$a_2^1 = \sigma(s_2^1) = \sigma(5.3) = 0.9950$$

Secondly, we repeat the operation to the output layer using the outputs of hidden layers $a_1^1$ and $a_2^1$ as inputs, we get:
$$s_1^2 = w_{11}^2 a_1^1 + w_{12}^2 a_2^1 + b_1^2 = 0.7 \times (0.9866) + 0.9 \times (0.9950) + 0.5 = 2.0862$$
$$y_1 = \sigma(s_1^2) = \sigma(2.0862) = 0.8896$$

Similarly, we can calculate the output value of the second neuron in the output layer:
$$s_2^2 = w_{21}^2 a_1^1 + w_{22}^2 a_2^1 + b_2^2 = 0.8 \times (0.9866) + 0.1 \times (0.9950) + 0.5 = 1.3888$$
$$y_2 = \sigma(s_2^2) = \sigma(1.3888) = 0.8004$$

In this example, we use the squared error function as the loss functionso the total error is to sum all the output errors together.

$$L = \frac{1}{2}[(o_1 - y_1)^2 + (o_2 - y_2)^2]$$

Since we only care about the derivative of the loss function with respect to the weights and bias, then we do not need to calculate the value of the loss function.

$$\frac{\partial L}{\partial y_1} = y_1 - o_1$$

$$\frac{\partial L}{\partial y_2} = y_2 - o_2$$

The next step is to backpropagate through the network to compute all the partial derivatives with respect to the parameters of weights and bias. In this phase, we will begin by calculating the derivatives in the output layer with respect to $w_{11}^2$, $w_{21}^2$, $w_{12}^2$, $w_{22}^2$. By applying the chain rule and considering $w_{11}^2$ we get:

$$\frac{\partial L}{\partial w_{11}^2} = \frac{\partial L}{\partial y_1} \cdot \frac{\partial y_1}{\partial y_2} \cdot \frac{\partial s_1^2}{\partial w_{11}^2} = (y_1 - o_1)(y_1(1 - y_1))a_1^1$$

$$= (0.8896 - 0.1) \times (0.8896 \times (1 - 0.8896)) \times (0.9866)$$

$$= 0.0765$$

where $\frac{\partial L}{\partial y_1} = (y_1 - o_1)$, $\frac{\partial y_1}{\partial s_1^2} = y_1(1 - y_1)$, $\frac{\partial s_1^2}{\partial w_{11}^2} = a_1^1$. Similarly, we can obtain other derivatives.

$$\frac{\partial L}{\partial w_{12}^2} = \frac{\partial L}{\partial y_1} \cdot \frac{\partial y_1}{\partial s_1^2} \cdot \frac{\partial s_1^2}{\partial w_{12}^2} = (y_1 - o_1)(y_1(1 - y_1))a_2^1$$

$$= (0.8896 - 0.1) \times (0.8896 \times (1 - 0.8896)) \times (0.9950)$$

$$= 0.0772$$

$$\frac{\partial L}{\partial w_{21}^2} = \frac{\partial L}{\partial y_2} \cdot \frac{\partial y_2}{\partial s_2^2} \cdot \frac{\partial s_2^2}{\partial w_{21}^2} = (y_2 - o_2)(y_2(1 - y_2))a_1^1$$

$$= (0.8004 - 0.05) \times (0.8004 \times (1 - 0.8004)) \times (0.9866)$$

$$= 0.1183$$

$$\frac{\partial L}{\partial w_{22}^2} = \frac{\partial L}{\partial y_2} \cdot \frac{\partial y_2}{\partial s_2^2} \cdot \frac{\partial s_2^2}{\partial w_{22}^2} = (y_2 - o_2)(y_2(1 - y_2))a_2^1$$

$$= (0.8004 - 0.05) \times (0.8004 \times (1 - 0.8004)) \times (0.9950)$$

$$= 0.1193$$

$$\frac{\partial L}{\partial b_1^2} = \frac{\partial L}{\partial y_1} \cdot \frac{\partial y_1}{\partial s_1^2} \cdot \frac{\partial s_1^2}{\partial b_1^2} = (y_1 - o_1)(y_1(1 - y_1))(1)$$

$$= (0.8896 - 0.1) \times (0.8896 \times (1 - 0.8896)) \times (1)$$

$$= 0.0775$$

$$\frac{\partial L}{\partial b_2^2} = \frac{\partial L}{\partial y_2} \cdot \frac{\partial y_2}{\partial s_2^2} \cdot \frac{\partial s_2^2}{\partial b_2^2} = (y_2 - o_2)(y_2(1 - y_2)) \times (1)$$

$$= (0.8004 - 0.05) \times (0.8004 \times (1 - 0.8004)) \times (1)$$

$$= 0.1199$$

Now we can proceed with backpropagation to the hidden layer to calculate the partial derivatives of the parameters connecting the input layer to the hidden layer, i.e., $w_{11}^1$, $w_{21}^1$, $w_{22}^1$, $w_{13}^1$, $w_{23}^1$. Starting with $\dfrac{\partial L}{\partial w_{11}^1}$, $\dfrac{\partial L}{\partial w_{12}^1}$, $\dfrac{\partial L}{\partial w_{13}^1}$, $\dfrac{\partial L}{\partial b_1^1}$ since they all flow through $a_1^1$. We can use equation (5.2) and equation (5.3) to obtain the results as:

$$\begin{aligned}\dfrac{\partial L}{\partial w_{11}^1} &= \delta_1^1 x_1 \\ &= \sigma(s_1^1)(1-\sigma(s_1^1))(\delta_1^2 w_{11}^2 + \delta_2^2 w_{21}^2)x_1 \\ &= a_1^1(1-a_1^1)[(y_1-o_1)y_1(1-y_1)(0.7)+[(y_2-o_2)y_2(1-y_2)(0.8)]x_1 \\ &= 0.9866 \times (1-0.9866) \times [(0.7896) \times (0.0983) \times (0.7)+(0.7504) \times (0.1598) \times \\ &\quad (0.8)] \times (1) \\ &= 0.0132 \times 0.1502 \times 1 \\ &= 0.0020\end{aligned}$$

$$\begin{aligned}\dfrac{\partial L}{\partial w_{12}^1} &= \delta_1^1 x_2 \\ &= \sigma(s_1^1)(1-\sigma(s_1^1))(\delta_1^2 w_{11}^2 + \delta_2^2 w_{21}^2)x_2 \\ &= a_1^1(1-a_1^1)[(y_1-o_1)y_1(1-y_1)(0.7)+[(y_2-o_2)y_2(1-y_2)(0.8)]x_2 \\ &= 0.9866 \times (1-0.9866) \times [(0.7896) \times (0.0983) \times (0.7)+(0.7504) \times (0.1598) \times \\ &\quad (0.8)] \times (4) \\ &= 0.0132 \times 0.1502 \times 4 \\ &= 0.0079\end{aligned}$$

$$\begin{aligned}\dfrac{\partial L}{\partial w_{13}^1} &= \delta_1^1 x_3 \\ &= \sigma(s_1^1)(1-\sigma(s_1^1))(\delta_1^2 w_{11}^2 + \delta_2^2 w_{21}^2)x_3 \\ &= a_1^1(1-a_1^1)[(y_1-o_1)y_1(1-y_1)(0.7)+[(y_2-o_2)y_2(1-y_2)(0.8)]x_3 \\ &= 0.9866 \times (1-0.9866) \times [(0.7896) \times (0.0983) \times (0.7)+(0.7504) \times (0.1598) \times \\ &\quad (0.8)] \times (5) \\ &= 0.0132 \times 0.1502 \times 5 \\ &= 0.0099\end{aligned}$$

$$\begin{aligned}\dfrac{\partial L}{\partial b_1^1} &= \delta_1^1 \\ &= \sigma(s_1^1)(1-\sigma(s_1^1))(\delta_1^2 w_{11}^2 + \delta_2^2 w_{21}^2) \\ &= a_1^1(1-a_1^1)[(y_1-o_1)y_1(1-y_1)(0.7)+[(y_2-o_2)y_2(1-y_2)(0.8)] \\ &= 0.9866 \times (1-0.9866) \times [(0.7896) \times (0.0983) \times (0.7)+(0.7504) \times (0.1598) \times (0.8)] \\ &= 0.0132 \times 0.1502 \\ &= 0.0047\end{aligned}$$

Similarly, we can calculate the partial derivatives of $\dfrac{\partial L}{\partial w_{21}^1}$, $\dfrac{\partial L}{\partial w_{22}^1}$, $\dfrac{\partial L}{\partial w_{23}^1}$, $\dfrac{\partial L}{\partial b_1^1}$, $\dfrac{\partial L}{\partial b_2^1}$.

$$\begin{aligned}\dfrac{\partial L}{\partial w_{21}^1} &= \delta_2^1 x_1 \\ &= \sigma(s_2^1)(1-\sigma(s_2^1))(\delta_1^2 w_{12}^2 + \delta_2^2 w_{22}^2)x_1\end{aligned}$$

$$=a_2^1(1-a_2^1)[(y_1-o_1)y_1(1-y_1)(0.7)+[(y_2-o_2)y_2(1-y_2)(0.8)]x_1$$
$$=0.9950\times(1-0.9950)\times[(0.7896)\times(0.0983)\times(0.9)+(0.7504)\times(0.1598)\times$$
$$(0.1)]\times(1)$$
$$=0.0049\times0.0818\times1$$
$$=0.0004$$

$$\frac{\partial L}{\partial w_{22}^1}=\delta_2^1 x_2$$
$$=\sigma(s_2^1)(1-\sigma(s_2^1)(\delta_1^2 w_{12}^2+\delta_2^2 w_{22}^2)x_2$$
$$=a_2^1(1-a_2^1)[(y_1-o_1)y_1(1-y_1)(0.7)+[(y_2-o_2)y_2(1-y_2)(0.8)]x_2$$
$$=0.9950\times(1-0.9950)\times[(0.7896)\times(0.0983)\times(0.9)+(0.7504)\times(0.1598)\times$$
$$(0.1)]\times(4)$$
$$=0.0049\times0.0818\times4$$
$$=0.0016$$

$$\frac{\partial L}{\partial w_{23}^1}=\delta_2^1 x_3$$
$$=\sigma(s_2^1)(1-\sigma(s_2^1)(\delta_1^2 w_{12}^2+\delta_2^2 w_{22}^2)x_3$$
$$=a_2^1(1-a_2^1)[(y_1-o_1)y_1(1-y_1)(0.7)+[(y_2-o_2)y_2(1-y_2)(0.8)]x_3$$
$$=0.9950\times(1-0.9950)\times[(0.7896)\times(0.0983)\times(0.9)+(0.7504)\times(0.1598)\times$$
$$(0.1)]\times(5)$$
$$=0.0049\times0.0818\times5$$
$$=0.0020$$

$$\frac{\partial L}{\partial b_2^1}=\delta_2^1$$
$$=\sigma(s_2^1)(1-\sigma(s_2^1)(\delta_1^2 w_{12}^2+\delta_2^2 w_{22}^2)$$
$$=a_2^1(1-a_2^1)[(y_1-o_1)y_1(1-y_1)(0.9)+[(y_2-o_2)y_2(1-y_2)(0.1)]$$
$$=0.9950\times(1-0.9950)\times[(0.7895)\times(0.90983)\times(0.9)+(0.7504)\times(0.1598)\times(0.1)]$$
$$=0.0049\times0.0818$$
$$=0.0004$$

Up to now, we have all the derivatives and finish one iteration after updating all the weights and bias parameters (I leave this to you as an EXERCISE). We set the learning rate $\eta$ equals 0.01. These steps should be repeated many times until the loss function converges some small values and the parameter becomes stable. Obviously, it is tedious and even impossible to go through all these calculations manually. Nowadays, we would rather employ a machine learning package that is demonstrated in the next section.

### 5.4.5　Choosing the right optimization algorithm

As presented in the BP step, for the training of a neural network, the optimization algorithm is used to determine the optimal weights for the model. We only employ gradient descent to illustrate the BP process, in practice, there are other alternatives to gradient descent in use including SGD, SGD with momentum, RMSProp, RMSProp with momentum,

AdaDelta, and Adam. Adaptive optimization methods such as Adam or RMSProp perform well during the process of training, but they were reported to have poor performance at later stages compared to SDG. Actually, all these optimizers have their own pros and cons (see TABLE 5-1), there is currently no consensus on a single best optimization algorithm that has adaptive capacity to all situations, hence, the choice of which algorithm to use depends largely on the user's familiarity with the algorithm.

TABLE 5-1  Pros and cons of some popular optimization algorithms

| Optimization Algorithm | Property | Pros | Cons |
| --- | --- | --- | --- |
| Gradient descent | Solve the optimal value along the direction of the gradient descent. The method converges at a linear rate | The solution is global optimal when the objective function is convex | In each parameter update, gradients of total samples need to be calculated, so the calculation cost is high |
| SGD | The update parameters are calculated using a randomly sampled mini-batch. The method converges at a sublinear rate | The calculation time for each update does not depend on the total number of training samples, and a lot of calculation cost is saved | It is difficult to choose an appropriate learning rate, and using the same learning rate for all parameters is not appropriate. The solution may be trapped at the saddle point in some cases |
| AdaGrad | The learning rate is adaptively adjusted according to the sum of the squares of all historical gradients | In the early stage of training, the cumulative gradient is smaller, the learning rate is larger, and the learning speed is faster. The method is suitable for dealing with sparse gradient problems. The learning rate of each parameter adjusts adaptively | As the training time increases, the accumulated gradient will become larger and larger, making the learning rate tend to zero, resulting in ineffective parameter updates. A manual learning rate is still needed. It is not suitable for dealing with non-convex problems |
| Adam | Combine the adaptive methods and the momentum method. Use the first-order moment estimation and the second-order moment estimation of the gradient to dynamically adjust the learning rate of each parameter. Add the bias correction | The gradient descent process is relatively stable. It is suitable for most non-convex optimization problems with large datasets and high dimensional space | The method may not converge in some cases |

continued

| Optimization Algorithm | Property | Pros | Cons |
| --- | --- | --- | --- |
| ADMM | The method solves optimization problems with linear constraints by adding a penalty term to the objective and separating variables into sub-problems which can be solved iteratively | The method uses the separable operators in the convex optimization problem to divide a large problem into multiple small problems that can be solved in a distributed manner. The framework is practical in most largescale optimization problems | The original residuals and dual residuals are both related to the penalty parameter whose value is difficult to determine |
| RMSProp | The method maintains a moving (discounted) average of the square of gradients and divides the gradient by the root of this average | RMSProp's adaptive learning rate usually prevents the learning rate decay from diminishing too slowly or too fast | RMSProp uses more memory for a given batch size than stochastic gradient descent and Momentum, but less than Adam |

## 5.5  Build a Neural Network in Python Keras

In this section, we will move to the implementation of an NN for classification by using the Python Keras framework. Keras is a deep learning API written in Python, running on top of the machine learning platform TensorFlow. It was developed with a focus on enabling fast experimentation. Keras contains numerous implementations of commonly used neural-network building blocks such as layers, objectives, activation functions, optimizers, and a host of tools to make working with image and text data easier to simplify the coding necessary for writing deep neural network code. Moreover, in addition to standard neural networks, Keras has support for convolutional and recurrent neural networks. It supports other common utility layers like dropout, batch normalization, and pooling. Before starting to code, make sure you have installed Keras on your computer. Or else, you can use the following code to install it: **pip install Keras.**

Now let's get started. To build a neural network we follow the steps as follows.

**Step 1**: Data preparation

In this example, we are going to build a neural network to tackle a classification problem that has two inputs and four outputs. The data used here is generated by the following code. Note here, we first should import some packages from Keras, which are necessary to define the function and classes.

```
# packages import
import keras
from keras.models import Sequential
from keras.layers import Dense
from keras.utils import to_categorical
import matplotlib.pyplot as plt
import numpy as np
from sklearn.datasets import make_blobs

# Generate data
num_samples_total = 1000 # the size of the dataset
training_split = 250
cluster_centers = [(15,0), (15,15), (0,15), (30,15)]
num_classes = len(cluster_centers)
loss_function_used = 'categorical_crossentropy' # the loss function used in the NN model
X, targets = make_blobs(n_samples = num_samples_total, centers = cluster_centers, n_features = num_classes, center_box= (0, 1), cluster_std = 1.5)

# plot the dataset
plt.scatter(X_training[:,0], X_training[:,1])
plt.title('Nonlinear data')
plt.xlabel('X1')
plt.ylabel('X2')
plt.show() # the result is shown in FIGURE 5-13.
```

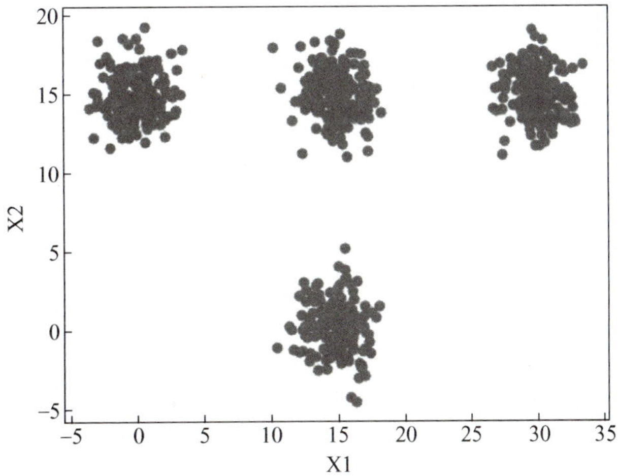

**FIGURE 5-13　The result of the dataset**

After dataset generation, its targets are converted into one-hot encoded vectors that are compatible with categorical crossentropy loss. Finally, we make the split between training and testing data.

```
# define the Training and testing size
X_training = X[training_split:,:] # training size (750) of input data
X_testing = X[:training_split,:] # testing size (250) of test data
categorical_targets = to_categorical(targets) # convert to one-hot vectors
Targets_training = categorical_targets[training_split:]
Targets_testing= ategorical_targets[:training_split].astype(np.integer)
```

**Step 2:** Keras model definition

Models in Keras are defined as a sequence of layers. We use the simple *Sequential* class to config the model rather than the model class. For simplicity, the implemented model has three layers with the final layer as the output layer.

```
# define the model
feature_vector_length= len(X_training [0])
input_shape = (feature_vector_length,) # the dataset has 2 features
model = Sequential ( [
    Dense (16, input_shape= input_shape, activation= 'relu')
    Dense (8, activation= 'relu'),
    Dense (num_classes, activation= 'softmax'), # num_classes= 4
])
```

Note that the line of code that adds the first Dense layer defines two things, i. e., the input layer and the first hidden layer which has 16 nodes. Once the input shape is specified, Keras will automatically infer the shapes of inputs for later layers. The last layer is a softmax output layer with 4 nodes, one for each cluster. So this is a very simple and densely-connected network.

**Step 3:** Compile Keras model

To compile the model, we should config the loss function, the optimizer used in the model, and some metrics to report the model performance.

```
# compile the model
model.compile(
    optimizer= 'adam', # The optimization algorithm to find the
                      # optimal weights and bias.
    loss= 'categorical_crossentropy', # loss function  metrics= ['accuracy'],)
```

**Step 4:** Fit Keras model

After model compilation, it is the time to fit or train the model on our generated dataset by calling the fit() function. Similarly, some parameters should be specified such as the number of epochs (iterations over the entire dataset), the batch size (number of samples per gradient update), the validation set. We fit the training data for 30 epochs with a batch size of 5.20% of the training data which is used for validation and all output is shown on screen with verbosity mode set to True (1).

```
# training the model
Model_fit = model.fit(X_training, Targets_training, epochs= 30, batch_size= 5,
verbose= 1, validation_split= 0.2)
```

**Step 5**: Evaluate Keras model

Once we finished the training of the model, the network can be evaluated by using the evaluate() function, which will return a list with two values: the loss and the accuracy of the model on the testing dataset.

```
# evaluate the model after training
test_results = model.evaluate(X_testing, Targets_testing, verbose= 1)
print(f'Test results - Loss: {test_results[0]} - Accuracy: {test_results[1]*
100}% ')
```

The outcome is shown as:

```
Test results - Loss: 0.0064070962513796985 - Accuracy: 100.0%
```

The model gets a 0.0064 test loss and 100% test accuracy. This is a perfect model, however, it is more complex in practice and usually can not obtain such an extremely good accuracy.

**Step 6**: Make predictions

The final step is to apply the trained model to predict on new data. Here we just pretend the training dataset is a new dataset we have never met before.

```
# make class predictions with thetraining dataset
predictions = model.predict_classes(X)
# summarize the first 5 cases
for i in range(5):
        print('% s = > % d (expected % d)' % (X[i].tolist(), predictions[i],
        targets[i]))
```

The results show that:

```
[29.926556091899926, 14.722438031572905] = > 3 (expected 3)
[16.065330498941137, 14.186734269367513] = > 1 (expected 1)
[15.440041013267932, - 0.5256789465920577] = > 0 (expected 0)
[14.182878084165683, 15.999072172462132] = > 1 (expected 1)
[- 0.25949566600903495, 15.33296040033949] = > 2 (expected 2)
```

Alternatively, the model can be stored on your disk for future applications to other new datasets for prediction. This can be done by means of the function of model: save_weights ('model.ws') and model.load_weights ('model.ws').

## 5.6 Deep Neural Networks

So far, all traditional or "plain vanilla" ANNs we have discussed should be called "shallow" networks that only have one hidden layer of neurons between inputs and outputs layers. Actually, there are a variety of ANN architectures for approaching many different complex problems. A Deep Neural Network (DNN) can have two or more "hidden layers" of neurons that process inputs. No doubt that deep learning is currently one of the most active research topics in the machine learning field. If you are interested in learning more about neural networks and algorithms for deep learning, I recommend you read the published paper on the introduction and overview. In this part, let's only take a brief look at two other popular DNN architectures, i.e., convolutional neural networks(CNN) and recurrent neural networks(RNN).

### 5.6.1 Convolutional Neural Networks (CNNs)

Convolutional Neural Networks (CNNs or ConvNets) are famous for their outstanding performance in areas of computer vision and image recognition. CNNs have proven successful in identifying images of faces, objects and traffic signs in addition to powering vision in robots and autonomous vehicles. CIFAR-10 is a dataset widely used for benchmarking computer vision algorithms in the field of machine learning, see FIGURE 5-14.

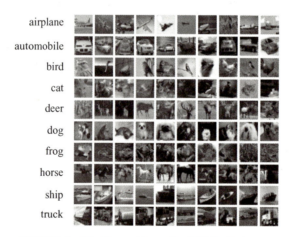

FIGURE 5-14　Images from the CIFAR-10 dataset

CNN employs a mathematical operation called convolution that is a specialized kind of linear operation. A CNN first performs convolution operations, which involve scanning the image, analyzing a small part of it each time, and creating a feature map with probabilities that each feature belongs to the required class (in a simple classification example). The second step is pooling, which reduces the dimensionality of each feature while maintaining its most important information. The max-pooling is the most common function, in which the highest value is taken from each pixel area scanned by the CNN, shown in FIGURE 5-15. The most common form is a pooling layer with filters of size $2 \times 2$ applied with a stride of 2 downsamples every depth slice in the input by 2 along both width and height, discarding 75% of the activations. Note the pooling operations can also perform functions other than the max-pooling, such as average pooling or even L2-norm pooling.

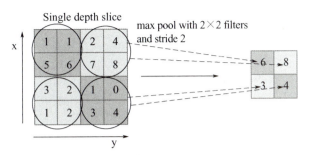

**FIGURE 5 – 15  The max pooling operation with a stride of 2 and 2×2 filters**

Finally, when the features are at the right level of granularity, it creates a fully-connected neural network that analyzes the final probabilities and decides which class the image belongs to. The final step can also be used for more complex tasks, such as generating a caption for the image. Typically, there are four main operations in the CNNs shown in FIGURE 5 – 16.

- Convolution.
- Non Linearity (ReLU).
- Pooling or Sub Sampling.
- Classification (Fully Connected Layer).

The convolution layers play the role of feature extractor by employing convolution operation on different features of the input. ReLU is a non-linear operation (see FIGURE 5 – 8) used to remove negative values from an activation map by setting them to zero without affecting the receptive fields of the convolution layer. The pooling is a form of non-linear down-sampling, serves to progressively reduce the spatial size of the representation, to reduce the number of parameters, memory footprint and amount of computation in the network, and hence to also control overfitting. Finally, after several convolutional and pooling layers, fully connected layers are used as a traditional Multi-Layer Perceptron that employs a softmax activation function in the output layer to decide which class the image belongs to.

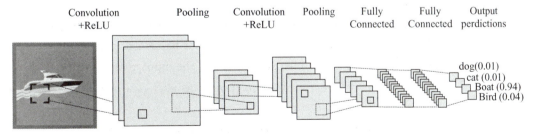

**FIGURE 5 – 16  A typical CNN architechture**

In summary, a CNN is a sequence of layers. Every layer of a CNN transforms one volume of activations to another through a differentiable function. Distinct types of layers, both locally and completely connected, are stacked to form a CNN architecture. CNNs give the best performance in pattern/image recognition problems and even outperform humans in certain cases. There are several other popular architectures in the field of CNNs, such as:

- LeNet.
- AlexNet.
- ZF Net.
- GoogLeNet.
- VGGNet.
- ResNet.
- CAPSNet.

These networks still remain at the forefront of commercial applications of deep learning today.

### 5.6.2 Recurrent Neural Networks (RNN)

A RNN is a deep learning algorithm derived from feedforward NN to process sequential data, such as natural language, speech recordings, video, audio, or multiple events that occur one after another. The basic RNN ( also called vanilla RNN) is a network of neuron-like nodes organized into successive layers. In a given layer, each node is connected with a directed (one-way) connection to every other node in the next successive layer (FIGURE 5 – 17). So it has a "memory" that captures historical information about what has been calculated so far.

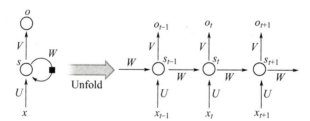

FIGURE 5 – 17  Unfolded basic RNN involved in its forward computation

- Inputs: $x_t$ is the input at time step $t$.
- Hidden output: $s_t$ is the hidden state at time step $t$. It's the memory part of an RNN, used to capture the historical information that happened in previous steps, it can be calculated by using the hidden state $s_{t-1}$ and the input at the current time step ($x_t$), i.e., $s_t = f(Ux_t + Ws_{t-1})$. The function $f$ usually is a nonlinearity such as $tanh$ or $ReLU$.
- Outputs: $o_t$—this is the model's prediction at time step $t$, $o_t = $softmax$(Vs_t)$.

For the backpropagation in RNN, it is called Backpropagation Through Time (BPTT). Indeed, BPTT uses the chain rule to go back from the latest step to the previous step, and then to the next-previous. Each time using gradient descent to discover the best weights for each neuron and for the hidden state function. These parameters define how much of the information from the previous steps should be carried forward to each subsequent step.

RNN has been used in many fields including language modeling and text generation, machine translation, speech recognition, time-series anomaly detection, and video tagging, etc. Furthermore, other sophisticated types of RNNs have been developed to tackle some of the drawbacks of the vanilla RNN model such as Bidirectional RNNs, Deep (Bidirectional) RNNs, and LSTM networks and GRU (Gated Recurrent Units). LSTM (Long Short Term Memory networks) can be viewed as a special type of RNN, designed to address the issue of vanishing gradients through a gating mechanism. The output of the first layer function is the input to the second layer and that output is the input to the third layer and so on. The length of the chain gives the depth of the model, and this is also an intuitive explanation of the term 'deep' used in deep learning. There are three gates used to regulate information propagation into and out of the unit of an LSTM: forget gate ($F_t$), input gate ($I_t$), and output gate ($O_t$). The structure of an LSTM unit in FIGURE 5-18 Among which, the forget gate determines what information should be kept or dropped out from prior steps, the input gate is responsible for what information is important to add to the unit state from the current step, the output gate performs the task of extracting useful information from the current cell state to be presented as output.

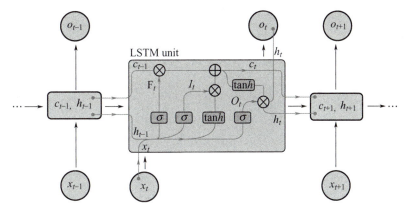

**FIGURE 5-18  The structure of an LSTM unit**

# EXERCISES

1. Please try to create a 4-layer ANN with one input layer of 3 neurons, one output layer of 2neurons, and 2 hidden layers that have 3 nodes in each layer.

2. If you have a perceptron shown below, what will the output be when the input vector ($x_1$, $x_2$) is (1, 2), and what if the given input data is (1, 0)? When we replace the bias weight value with $-2.0$, Please give your answer with detailed steps. The activation function used is supposed to be the Heaviside step function.

3. With the NN structure in Figure 5-11, please use BP algorithm update the weights and bias parameters when the input vector is changed to [1, 2, 3], the outputs become [0.1, 0.9] after one iteration. The learning rate is set to 0.01. And implement the NN with Keras to get the optimal weights and bias.

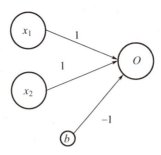

4. Programming task: please use what you have learned to build your own CNN and train it with MNIST dataset, and give your prediction results (accuracy) (data source: https://keras.io/api/datasets/mnist/).

5. Find at least 3 techniques to deal with overfitting problems of ANN.

6. Perform a literature review concerning recent studies on Deep Learning Techniques in applications of business management. (no less than 6000 words)

本章配套资源

# Chapter 6
# Applications of Data Mining in Management Decision-making

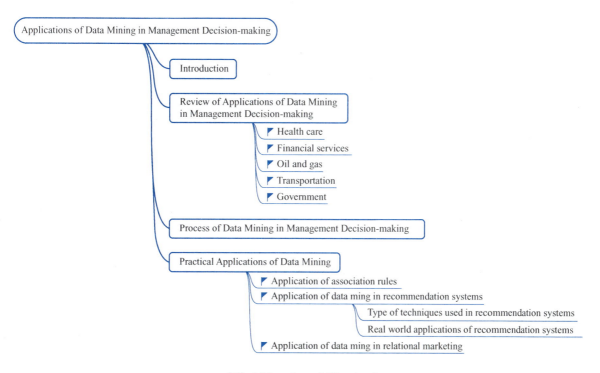

**Mind Mapping of Chapter 6**

众所周知，管理决策对企业很重要，但是做出有效的决策却是一个复杂的过程。决策策略并不总是通过使用传统模型来产生预期的结果，特别是在动态数据驱动的环境中。在大数据时代，数据挖掘技术特别是机器学习、大数据分析、云计算等技术，正在改变着全球几乎所有行业的业务格局。海量商业交易数据库的可用性提供了有关客户如何使用服务或购买产品的准确信息。而数据挖掘可以帮助企业更好地了解其业务和市场，并及时做出业务决策，这是因为它能够从数据库模式中识别。

本章重点阐述数据挖掘在管理决策中的应用。首先，快速回顾数据挖掘技术在管理决策中的应用情况；其次，介绍数据挖掘的通用框架或过程；最后，举例说明了一些流行的数据挖掘技术（如关联规则、分类等）在推荐系统、市场营销等业务领域的实际应用。

## 6.1  Introduction

As is known, decision-making for management is important for an enterprise, however, to make efficient decisions is a complex process that is influenced by many factors. Decision strategies do not always succeed in producing desired outcomes by using traditional models, particularly in dynamic data-driven environments.

In the era of big data, data mining technologies, especially advanced techniques such as machine learning, Big Data analytics and cloud computing are transforming the business patterns of almost all industries across the world. The availability of massive databases of business transactions provides accurate information on how customers make use of services or purchase products. Data mining has potential applications to help an enterprise better understand his business and market and make timely business decisions due to its capability to learn from the databases and identify patterns, connections and insights.

Therefore, after finishing the discussion of the main techniques of data mining in previous chapters, in this chapter we will make our focus primarily on the applications of data mining in management decision-making. In the sections below, firstly a quick review of the applications of data mining techniques in management decision-making. Then the general data mining framework or process is introduced. Lastly, examples are presented on practical applications of some popular data mining techniques in addressing issues related to making business decisions in the area of recommendation systems and market intelligence.

## 6.2  Review of Applications of Data Mining in Management Decision-making

In many different industries, data mining is exerting profound impacts on a large scale of areas from financial services to advertising and marketing, and becoming a fundamental tool to assist leaders (decision-makers) to make better decisions backed by data instead of intuition or experience even though this is important in some situations.

For now, more companies tend to set up data mining frameworks to analyze business data and to automate well-structured decisions. For example, when faced with a choice, problem, or work item to be handled, data mining can report to decision-makers for what their typical answers have been in the past for that specific situation based on its learning techniques and then provide intelligent solutions for decision-making. Furthermore, the internet and the WWW have made the process of collecting data more convenient and there are more affordable data storage solutions and greater computational processing power to support the data mining techniques. This will dramatically change the traditional process of making decisions, even the way one thinks about business. In the next 5 –10 years, data mining will be used to enable enterprises and organizations to optimize and improve a wide range of business processes to meet their digital transformation goals and then to gain an advantage over competitors.

Under such circumstances, decision-making assisted by data mining algorithms is having an increasing influence on our lives. From the perspective of business intelligence and analytics (BI&A), data mining includes not only data processing and analytical technologies, but business-centric practices and methodologies that can be applied to various high-impact fields such as e-commerce, market intelligence, e-government, healthcare, and security. Moreover, some emerging data analytics have been developed and got much research interest from the academic community. The evolution of BI&A, applications and emerging techniques are shown in FIGURE 6 – 1.

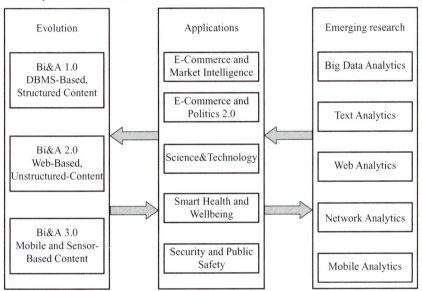

FIGURE 6 – 1　Evolution of business intelligence and analytics (BI&A), applications and emerging techniques

More application examples can be found in the fields of healthcare, manufacturing control, financial fraud detection, and customer relation management. The following are some typical application areas of data mining.

- Health care

Data mining is a fast-growing trend in the health care industry on account of the advent of wearable devices and sensors that can use data to assess a patient's health in real-time. The technology can also help measure the effectiveness of certain treatments and assist medical experts to identify trends or red flags that may lead to improved diagnoses and treatment.

- Financial services

Banks and other businesses in the financial industry use data mining technologies for two key purposes: to identify important insights in data, and prevent fraud. The insights can identify investment opportunities, or help investors know when to trade. Data mining can also identify clients with high-risk profiles, or use cyber surveillance to pinpoint warning signs of fraud.

- Oil and gas

With the proliferation of resource exploitation in both traditional and unconventional basins, more upstream oil and gas industry activities appear in many regions. Management and use of data mining techniques have become critical for the industry and its stakeholders, including regulators and financiers. The integration of data mining analytics into the practice of petroleum engineering becomes essential to build a vision for the oil and gas industry to move toward data-driven decisions in the production and operations area. There are many cases using data mining in this industry such as finding new energy sources, analyzing minerals in the ground, predicting refinery sensor failure, streamlining oil distribution to make it more efficient and cost-effective.

- Transportation

With the surge of traffic data, applying data mining techniques to identify traffic patterns and trends is recently getting more and more attention from the decision-makers of transportation. Data mining techniques and machine learning methodologies have been used to address traffic management problems such as traffic jam, traffic safety, travel behavior modeling, transportation mode recognition, and ITS.

- Government

Government sectors such as public safety and municipal utilities have a particular demand for data mining techniques. They have multiple sources of data that can be mined for insights in regard to crime prevention, detecting fraud in government construction procurement, public goods management, e-government services and policy evaluation.

## 6.3 Process of Data Mining in Management Decision-making

FIGURE 6-2 presents the general process of a data mining system. Typically, it has six steps. But note that these steps can be customized based on the requirement of the project.

(1) Data acquisition (collect and gather dataset): The first step is to collect the relevant dataset. Usually, a subject matter expert identifies the features relevant to the problem.

(2) Data preparation (join, filter, cleanse data sets): This step is to clean the dataset to remove noise and handle missing data, the data is reduced (converted) to the desired format and then enable you to feed data to the selected training model.

(3) Model training and testing: The training data (typically 80%) is used to train and build the model and the model is evaluated by its performance on the testing data (typically 20%). A popular evaluation technique, known as, $k$-fold cross-validation, is usually used

to measure the performance of the model under consideration when the dataset has sufficient size.

(4) Model deployment: As you tried multiple models during the evaluation, you have to clean up everything and make sure that only one model is left now, with the optimum performance. Now it is time to publish/deploy the model.

(5) Diagnose issues and refine the system: Generally, the model does not meet the success criteria for the first time, then one should diagnose the issues and refine the system. Diagnosing issues and refinement activities must be repeated until the system is performing at a level necessary for the intended application. If the corrective action is not taken at the time, this will lead to a reduction in prediction accuracy.

(6) Bring insights and values: After the completion of the above steps, the model is now able to guide better business decisions and more intelligent actions with minimal human intervention.

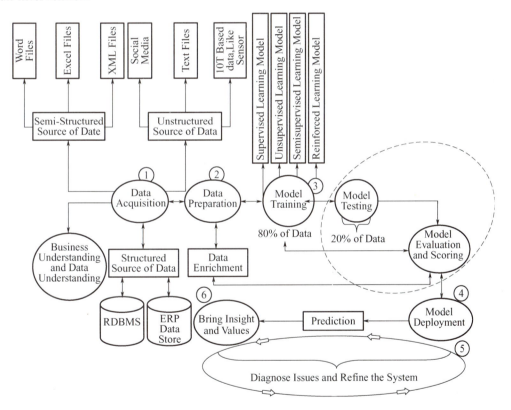

FIGURE 6 – 2  The general process of a data mining system

## 6.4 Practical Applications of Data Mining

In this section, the recommendation system and relational marketing are selected as the cases to make the practical applications of data ming more explicit.

### 6.4.1 Application of association rules

Because of its retail origins, association rule mining is often referred to as market basket analysis. It is used to explain patterns in data from seemingly independent information repositories, such as relational databases and transactional databases. These mined rules enable us to create pinpointed marketing proposals, special promotions, and winning formulas. They play an important part in customer analytics, market basket analysis, product clustering, catalog design, and store layout. With advances in data mining, association rules can be used in a wider breadth of application cases. Below are a few real-world use cases for association rules.

(1) Medicine. Doctors can use association rules to help diagnose patients. There are many variables to consider when making a diagnosis, as many diseases share symptoms. By using association rules and machine learning-fueled data analysis, doctors can determine the conditional probability of a given illness by comparing symptom relationships in the data from past cases. As new diagnoses get made, the model can adapt the rules to reflect the updated data.

(2) Retail. Retailers (especially online retailers) can collect data about purchasing patterns, recording purchase data as item barcodes are scanned by point-of-sale systems. The models can look for co-occurrence in this data to determine which products are most likely to be purchased together. The retailer can adjust marketing and sales strategy to take advantage of this information.

(3) User experience design. Developers can collect data on how consumers use a website they create. They can then use associations in the data to optimize the website user interface by analyzing where users tend to click and what maximizes the chance that they engage with a call to action.

(4) Entertainment. Based on the analysis of past user behavior data for frequent patterns, the association rules algorithm is used to recommend content that a user is likely to engage with, or organize content in a way that is likely to put the most interesting content for a given user.

### 6.4.2 Application of data ming in recommendation systems

Growing industries are targeting their products and advertisements for consumers based on data mining techniques. In particular, this is the case for e-commerce and retail companies, who are boosting sales by implementing recommender systems on their websites by leveraging the power of data.

A recommendation system (or a recommender platform/engine), aims to predict the "rating" or "preference" that a user would give to an item. Recommender systems are among the most popular applications of data mining today. They are utilized in a variety of areas including e-commerce, online advertisement, and financial service, etc. Some specific topics like restaurants and online dating are also popular application fields forthe recommendation system.

There are some potential benefits of using recommendation systems in business, such as improving the customer intention of products, improving cart value by helping customers find more of what they need, and enhancing engagement and delight. Therefore, the recommendation systems are important and valuable tools for these internet companies. The companies collect and uncover demographic data from customers and integrate them with information from previous purchases, product ratings, and customer behavior, then obtain knowledge to make decisions.

### 1. Type of techniques used in recommendation systems

Recommendation systems use a number of different techniques (FIGURE 6-3). The two typical approaches used in recommendation systems are collaborative filtering and content-based filtering (also known as the personality-based approach). Collaborative filtering (CF) techniques make predictions of what probably may interest a user based on his/her or other users' past behavior (items previously purchased or selected and/or numerical ratings given to those items). Content-based filtering (CBF) techniques rely on the pre-tagged characteristics of the products themselves to recommend additional items with similar attributes, ignoring contributions from other users.

A key advantage of the collaborative filtering approach is that it does not rely on machine analyzable content and therefore it is capable of accurately recommending complex items such as movies without the requirement of understanding the movies themselves. However, the collaborative filtering approaches often suffer from problems of cold start, scalability, and sparsity as below.

- Cold start: For a new user or item, there isn't enough data to make accurate recommendations.
- Scalability: In many environments used to make recommendations, there are millions of users and products. Thus, a large amount of computation power is often necessary to support the calculation of the algorithms.
- Sparsity: The number of items sold on major e-commerce online stores is extremely large. The most active users will only have rated a small subset of the overall database. Thus, even the most popular items have very few ratings.

Content-based filtering techniques are also facing main issues of limited content analysis, overspecialization and sparsity of data. To overcome some of the identified problems, hybrid filtering techniques have been proposed, which can implement a combination of two or more approaches in different ways to improve the performance of recommender systems.

Collaborative filtering can be classified into two types: a memory-based model and a model-based model. For model-based filtering techniques, many data mining algorithms have been applied to measure the similarity of users or items in recommender systems, such as clustering, association rules, Bayesian-networks model and Neural Networks (see FIGURE 6-3). We take the user-based nearest neighbor for illustration. Assume John is an active Netflix user and has not seen a video "A" yet. Thus the algorithm will perform as below.

(1) The similarity is calculated to find a set of users or nearest neighbors who have the same preferred items as John in the past and have rated video "A".

(2) The algorithm makes predict.

(3) Check for all the items John has not seen and recommends.

In essence, the CF algorithm generates a prediction for item $i$ by analyzing the rating for $i$ from neighbor users with higher similarity. For memory-based filtering techniques, it can be divided into two main sections: user-based and item-based filtering algorithms, and the item-based collaborative filtering is faster and more stable than user-based for a system where there are more users than items.

Once the rating of items is generated then we can provide the recommendations.

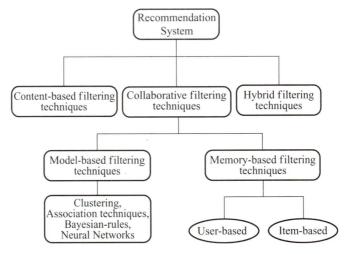

FIGURE 6-3　The techniques of recommendation systems

### 2. Real world applications of recommendation systems

Many famous internet companies are using recommendation systems. Here we just dictate how Netflix's Recommendations System works. Netflix is a platform that provides online movie and video streaming. Netflix Real-time data have more than 20 000 movies and shows and 2 million users. The Netflix recommendation system offers recommendations by matching and searching similar users' behavior and suggesting movies that share characteristics with films that users have rated highly. These recommendations can be applied to every user based on his/her unique profile. The workflow of the recommendation system used by Netflix is given in FIGURE 6-4.

The following insights are from the Help Center of Netflix and described in plain language and should be easier to understand for non-technical managers and execs.

Whenever Netflix service is visited, the recommendations system strives to help a user find a desired show or movie with minimal effort. The likelihood that the user will watch a particular title in the catalog of Netflix is calculated based on factors including:

(1) Interactions with the provide service (such as the user's viewing history and how the user rated other titles).

(2) Other members with similar tastes and preferences on our service.

(3) Information about the titles, such as the genre, categories, actors, etc.

Additionally, to best personalize the recommendations, Netflix also considers things such as the time of day, when and how long the user watches the movies; the device the user is watching Netflix on. All of these pieces of data are used as inputs to the developed recommendation algorithms. Note that the recommendation system does not include demographic information (such as age or gender) as part of the decision making process. In particular, when entering a search query, the top results returned are based on the actions of other members who have entered the same or similar queries. Generally, the collected data, algorithms, and computation systems continue to feed into each other to produce fresh recommendations to provide the user with a product.

A tutorial of the implementation of the Netflix Movie Recommendation System with Python can be found on the website: *https://github.com/towardsai/tutorials/tree/master/recommendation_system_tutorial*.

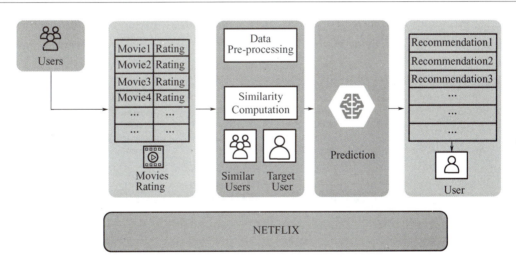

**FIGURE 6 – 4** The workflow of the recommendation system used by Netflix

### 6.4.3 Application of data ming in relational marketing

In today's changing landscape and contour of global marketing with a wide range of social media tools available, modern marketers are shifting their focus from transactional marketing to relational marketing. Relational marketing has become more important for enterprises that attempt to develop targeted and effective marketing campaigns. The goal of a relational marketing strategy is to initiate, strengthen, intensify and preserve over time the relationships between a company and its stakeholders represented primarily by its customers, and then to maintain a sustainable competitive advantage. Relational marketing strategies spread so fast due to the changes in the business environment which include the increased flow of information, available large amounts of data on customers' behavior, and the dramatic growth in the number of customers leading to greater complexity in the markets. To carry out relational marketing is not easy to do, a company should follow correct and careful approaches. What's more, it is not merely a collection of software applications or combinations of data mining techniques. In contrast, it is an integrated project where the various departments of an organization need to cooperate besides the adoption of business intelligence and data mining analytical tools.

Knowledge management for relational marketing can come from three major sources: customer knowledge from the retailer, consumer knowledge from market research, and integrated knowledge management system for marketing (FIGURE 6-5). The process of tightly integrating marketing decisions with the knowledge gained from knowledge discovery is known as knowledge-based relational marketing. There are some major areas of application of data mining for knowledge-based relational marketing. (a) Customer profiling. A customer profile is a model of the customer, based on which the marketer decides on the right strategies and tactics to meet the needs of that customer. (b) Deviation analysis. A deviation can be an anomaly fraud or a change. In the past, such deviations were difficult to detect in time to take corrective action, data mining tools provide powerful means such as neural networks for detecting and classifying such deviations. (c) Trend analysis. Trends are patterns that persist over some time and can be used for evaluating relational marketing programs or to forecast future sales.

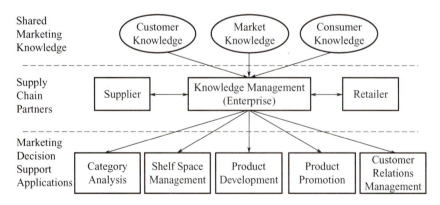

FIGURE 6-5　Integrated knowledge management system for marketing

The main phases of a relational marketing analysis proceed as shown in FIGURE 6-6. The first step is the exploitation of the data concerning each customer. Then, by using data mining techniques, it is expected to extract from those data the insights and the rules that allow market segments characterized by similar behaviors to be identified. For instance, a classification model can be used to generate a scoring system for customers according to their propensity of buying a service/product offered, and then the model directs a cross-selling recommendation specifically toward those customers for whom a high probability of acceptance is predicted by the model, thus maximizing the overall redemption of the marketing actions. Next, knowledge of customer profiles is used to design marketing actions which are then translated into promotional campaigns and generate in turn new information/knowledge to be used in the course of subsequent analyses. The corresponding decision-making process can be formally performed by appropriate optimization models. The cycle of marketing activities terminates with the execution of the planned campaign, with the subsequent gathering of information on the results and the redemption among the recipients.

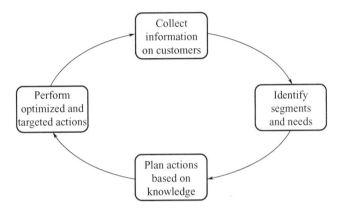

FIGURE 6-6  Main phases of a relational marketing analysis process

# EXERCISES

1. What is the general process of data mining in management decision-making? Do you have some different viewpoints on the process discussed in this chapter?

2. What do you think the most important step is for applications of data mining in practical problem approaching process?

3. Consider a bank head who wants to search new ways to increase revenues from its credit card operations. He has a vast data pool of customer information like age, gender, income, credit history, etc. He expects to check whether usage would double if fees were halved. What techniques can the head use to realize the goal?

4. Can you illustrate some negative effects of data mining applications in reality? What are your suggestions for mitigating these adverse impacts?

5. Gather at least three data mining business applications that you consider to be the most successful.

本章配套资源

# Appendix

## A1  Basic Steps for Data Mining

Following a structured approach to data mining helps you to maximize your chances of success in a data science project at the lowest cost. It also makes it possible to take up a project as a team, with each team member focusing on what they do best. However, this approach may not be suitable for every type of project or be the only way to do good data mining. The typical data mining process typically consists of six steps through which you can iterate, as shown below.

FIGURE A-1 presents the main steps and actions you might cover in data mining process. Each of these steps is introduced briefly.

(1) The first step of this process is setting a research goal. In particular, for a project, it should be explicitly understood by every practitioner and stakeholder.

(2) The second step is data retrieval. Data is the most critical part for any kind of data mining task. This step includes searching for suitable data and getting access to them from various data sources. This data is mostly in their raw format, which probably needs polishing and transformation before it becomes usable.

(3) Now that you have the raw data for model building preparation. This includes transforming the data from a raw form into a format that can be directly used in the models.

(4) The fourth step is data exploration. The goal of this step is to gain a deep understanding of the data such as patterns, correlations, and deviations based on visual and descriptive techniques.

(5) Model building is the most important part: It is time to gain insights or make predictions stated in your project charter. According to the goal, some techniques are used to build the prediction model.

(6) The last step of the data mining is presenting your results and automating the analysis if needed. One goal of a project is to help make better decisions. These results may convince the business and change the business process as expected, then improve your business efficiency and revenue.

Appendix | 145

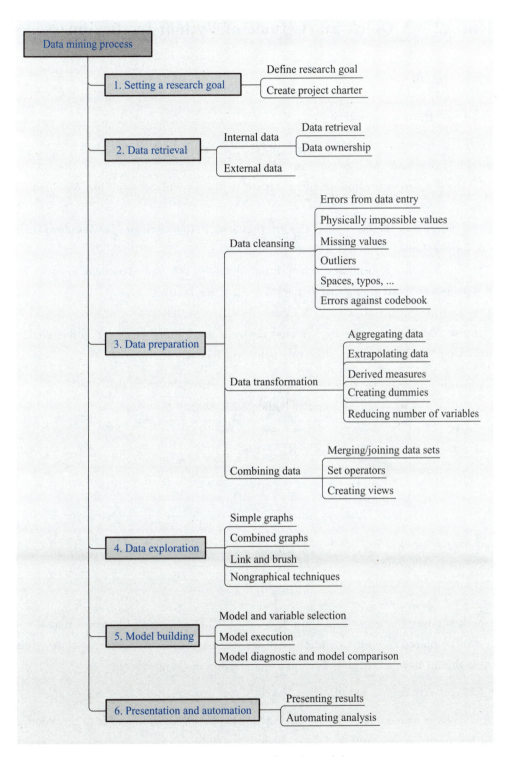

FIGURE A-1  The six steps of the data mining process

# A2   A Quick Start Guide of Python for Beginners

## A2.1   Install

### A2.1.1   Python installation

Here is a brief introduction on how to install Python 3 on Windows.

**Step 1**: Download the Python 3 Installer

(1) Open a browser window and navigate to the Download page for Windows at python. org.

(2) Underneath the heading at the top that says **Python Releases for Windows**, click on the link for the **Latest Python 3 Release-Python 3. x. x.**

(3) Scroll to the bottom and select either **Windows x86 – 64 executable installer** for 64-bit or **Windows x86 executable installer** for 32-bit. (See below)

**Step 2**: Run the Installer

After the download of the Python package, you can simply run it by double-clicking on the file. A dialog should appear like this below (FIGURE A – 2).

FIGURE   A – 2

Check the box that says Add Python 3. 7 to PATH as shown to ensure that the interpreter will be placed in your execution path. Then just click *Install Now*. A few minutes later you should have a working Python 3 installation on your system.

### A2.1.2   Anaconda installation

Anaconda is a free and open-source distribution of the Python and R programming languages for scientific computing (data science, machine learning applications, large-scale data processing, predictive analytics, etc.). It aims to simplify package management and deployment. Here we introduce how to install it on your windows system. The instructions are from the homepage of Anaconda (https://docs. anaconda. com/anaconda/install/

windows/).

(1) Download the Anaconda installer.

(2) Double click the installer to launch.

(3) Click Next.

(4) Read the licensing terms and click "I Agree".

(5) Select an install for "Just Me" unless you're installing for all users (which requires Windows Administrator privileges) and click Next.

(6) Select a destination folder to install Anaconda and click the Next button. See FIGURE A-3.

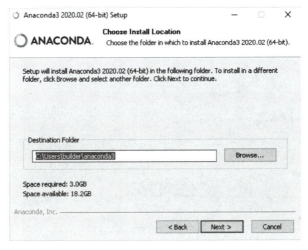

FIGURE A-3

(7) Choose whether to add Anaconda to your PATH environment variable. We recommend not adding Anaconda to the PATH environment variable, since this can interfere with other software. Instead, use Anaconda software by opening as the selection in FIGURE A-4 Anaconda Navigator or the Anaconda Prompt from the Start Menu.

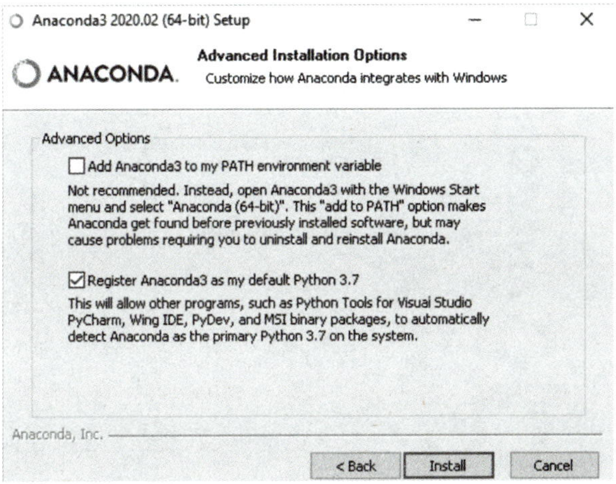

FIGURE A-4

(8) Choose whether to register Anaconda as your default Python. Unless you plan on installing and running multiple versions of Anaconda or multiple versions of Python, accept the default and leave this box checked.

(9) Click the Install button. If you want to watch the packages Anaconda is installing, click Show Details.

(10) Click the Next button.

(11) Optional: To install PyCharm for Anaconda, click on the link to https://www.anaconda.com/pycharm. Or to install Anaconda without PyCharm, click the Next button in FIGURE A – 5.

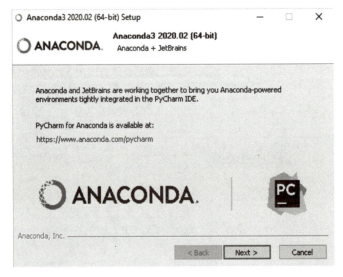

FIGURE　A – 5

(12) After a successful installation you will see "Thank you for installing Anaconda Individual Edition" dialog box, shown as FIGURE A – 6.

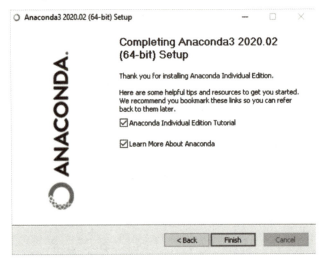

FIGURE　A – 6

## A2.2 Some Cheat-sheets you should know about Python

### A2.2.1 Numpy

Numpy is the core library of Python data science computing, providing support for large, multi-dimensional arrays and matrices, along with a large collection of high-level mathematical functions to operate on these arrays.

**Use the following statement to import the Numpy library:**

&gt;&gt;&gt;import numpy as np  #you can use other names instead of "np"

#### A2.2.1.1 Create array
&gt;&gt;&gt;a=np.array([1, 2, 3])
&gt;&gt;&gt;b=np.array([(1.5, 2, 3), (4, 5, 6)], dtype=float)
&gt;&gt;&gt;c=np.array([[(1.5, 2, 3), (4, 5, 6)], [(3, 2, 1), (4, 5, 6)]], dtype=float)

#### A2.2.1.2 Initialize placeholder
&gt;&gt;&gt;np.zeros((3, 4))                    #Create an array with value 0
&gt;&gt;&gt;np.ones((2, 3, 4), dtype=np.int16)   #Create an array with value 1
&gt;&gt;&gt;d=np.arange (10, 25, 5)              #Create an evenly spaced array with a step value of 5
&gt;&gt;&gt;np.linspace (0, 2, 9)                #Create an evenly spaced array with a number of samples of 9
&gt;&gt;&gt;e=np.full((2, 2), 7)                 #Create constant array with a result: array([[7, 7], [7, 7]])
&gt;&gt;&gt;f=np.eye (2)                         #Create a 2x2 identity matrix
&gt;&gt;&gt;np.random.random((2, 2))             #Create an array of random values
&gt;&gt;&gt;np.empty((3, 2))                     #Create empty array

#### A2.2.1.3 Input/Output

(1) Save and load files on your disk
&gt;&gt;&gt;np.save ('my_array', a)
&gt;&gt;&gt;np.savez ('array.npz', a, b)
&gt;&gt;&gt;np.load ('my_array.npy')

(2) Save and load text files
&gt;&gt;&gt;np.loadtxt("myfile.txt")
&gt;&gt;&gt;np.genfromtxt("my_file.csv", delimiter=',')
&gt;&gt;&gt;np.savetxt("myarray.txt", a, delimiter=" ")

(3) Type of data
&gt;&gt;&gt;np.int64      #Signed 64-bit integer
&gt;&gt;&gt;np.float32    #Standard double precision floating point
&gt;&gt;&gt;np.complex    #Complex number displayed as a 128-bit floating point number
&gt;&gt;&gt;np.bool       #Boolean values: True and False values

```
>>>np.object         # Python objects
>>>np.string_        # Fixed-length string
>>>np.unicode_       # Fixed-length unicode
```

### A2.2.1.4　Array information

```
>>>a.shape          # Array shape, several rows and columns
>>>len(a)           # Array length
>>>b.ndim           # Several-dimensional array
>>>e.size           # How many elements does the array have
>>>b.dtype          # type of data
>>>b.dtype.name     # The name of the data type
>>>b.astype(int)    # Data type conversion
```

### A2.2.1.5　Array calculation

(1) Arithmetic operations

```
>>>g=a-b                                   # Subtraction
   array([[−0.5, 0. , 0. ],
          [−3. , −3. , −3. ]])
>>>np.subtract(a, b)                       # Subtraction
>>>b + a                                   # addition
   array([[ 2.5, 4. , 6. ],
          [ 5. , 7. , 9. ]])
>>>np.add(b, a)                            # addition
>>>a / b                                   # division
   array([[0.66666667, 1. , 1. ],
          [ 0.25, 0.4, 0.5]])
>>>np.divide(a, b)                         # division
>>>a * b                                   # multiplication
   array([[1.5, 4. , 9. ],
          [4. , 10. , 18. ]])
>>>np.multiply(a, b)                       # multiplication
>>>np.exp(b)                               # power
>>>np.sqrt(b)                              # Square root
>>>np.sin(a)                               # Sine function
>>>np.cos(b)                               # Cosine function
>>>np.log(a)                               # Natural logarithm
>>>e.dot(f)                                # Dot product
   array([[ 7., 7.],
          [ 7., 7.]])
```

(2) Compare
```
>>>a==b                                    #Contrast value
  array([[False, True, True],
         [False, False, False]], dtype=bool)
>>>a<2                                     #Contrast value
  array([True, False, False], dtype=bool)
>>>np.array_equal(a, b)                    #Compare array
```
(3) Aggregate function
```
>>>a.sum()              #Array summary
>>>a.min()              #Array minimum
>>>b.max(axis=0)        #Maximum value of the array, by row
>>>b.cumsum(axis=1)     #Accumulated value of array elements
>>>a.mean()             #Average
>>>b.median()           #Medium number
>>>a.corrcoef()         #Correlation coefficient
>>>np.std(b)            #Standard deviation
```
(4) Array sort
```
>>>a.sort()             #Array sort
>>>c.sort(axis=0)       #Sort array by axis
```
(5) Subset
```
>>>a[2]
  3                     #Select the value corresponding to index 2
>>>b[1, 2]              #Select the value corresponding to row 1 and column 2
  6.0                   (equivalent to b[1][2])
```
(6) Slice
```
>>>a[0: 2]              #Select the value corresponding to in-
  array([1, 2])         dex 0 and 1
>>>b[0: 2, 1]           #Select the value of row 0 and row 1 in
  array([2., 5.])       column 1
>>>b[:1]                #Select all values in row 0 (equivalent
  array([[1.5, 2., 3.]])  to b[0: 1,: 1]
>>>c[1,…]               #Equivalent to [1,:,:]
  array([[[3., 2., 1.],
          [4., 5., 6.]]])
>>>a[ : : -1]
  array([3, 2, 1])      #Reverse array a
```
(7) Condition index
```
>>>a[a<2]
  array([1])            #Select all values in array a that are less than 2
```

(8) Fancy Index

```
>>>b[[1, 0, 1, 0], [0, 1, 2, 0]]              # Select the value corresponding to
  array([4. , 2. , 6. , 1.5])                   (1, 0), (0, 1), (1, 2) and (0, 0)
>>>b[[1, 0, 1, 0]][:, [0, 1, 2, 0]]
  array([[4. , 5. , 6. , 4. ],                  # Select a subset of the rows and
         [1.5, 2. , 3. , 1.5],                  columns of the matrix
         [4. , 5. , 6. , 4. ],
         [1.5, 2. , 3. , 1.5]])
```

### A2.2.1.6 Array operations

(1) Transpose array

```
>>>i=np.transpose(b)           # Transpose array
>>>i.T                          # Transpose array
```

(2) Change the shape of the array

```
>>>b.ravel()                    # Flatten the array
>>>g.reshape(3, -2)             # Change the shape of the array, but not the data
```

(3) Add or deletea value

```
>>>h.resize((2, 6))             # Return a new array of shape(2, 6)
>>>np.append(h, g)              # Append data
>>>np.insert(a, 1, 5)           # Insert dat
>>>np.delete(a, [1])            # delete data
```

(4) Merge array

```
>>>np.concatenate((a, d), axis=0)       # Concatenation array
  array([1, 2, 3, 10, 15, 20])
>>>np.vstack((a, b))
  array([[1. , 2. , 3. ],               # Stack arrays vertically in row di-
         [1.5, 2. , 3. ],               mensions
         [4. , 5. , 6. ]])
>>>np.r_[e,f]                           # Stack arrays vertically in row di-
                                        mensions
>>>np.hstack((e, f))
  array([[7., 7., 1., 0.],              # Stack arrays horizontally in col-
         [7., 7., 0., 1.]])             umn dimensions
>>>np.column_stack((a, d))
  array([[1, 10],                       # Create a stacked array with col-
         [2, 15],                       umn dimensions
         [3, 20]])

>>>np.c_[a,d]                           # Create a stacked array with col-
                                        umn dimensions
```

(5) Split array

```
>>>np.hsplit(a, 3)
[array([1]), array([2]), array([3])]
```
# Split the array vertically into 3 equal parts

```
>>>np.vsplit (c, 2)
[array([[ 1.5, 2. , 1. ],
        [ 4. , 5. , 6. ]]),
 array([[ 3. , 2. , 3. ],
        [ 4. , 5. , 6. ]])]
```
# Split the array horizontally into 2 equal parts

### A2.2.2 Pandas

Pandas is a Python library based on Numpy that provides easy-to-use data structures and data analysis tools for Python.

**Use the following statement to import the Pandas library:**

```
>>>import pandas as pd
```

#### A2.2.2.1 Series-Sequence

**One-dimensional array of any type of data**

```
>>>s=pd.Series([3,-5, 7, 4], index=['a', 'b', 'c', 'd'])
```

#### A2.2.2.2 DataFrame

(1) Two-dimensional arrays of different types of data

```
>>>data={'Country': ['Belgium', 'India', 'Brazil'],
'Capital': ['Brussels', 'New Delhi', 'Brasília'],
'Population': [11190846, 1303171035, 207847528]}
>>>df=pd.DataFrame (data, columns=['Country', 'Capital', 'Population'])
```

(2) Selection, Boolean index and setting value

By location

```
>>>df.iloc[[0], [0]]
   'Belgium'
>>>df.iat([0], [0])
   'Belgium'
```
# Select a value by row and column position

By label

```
>>>df.loc[[0], ['Country']]
'Belgium'
>>>df.at([0], ['Country'])
   'Belgium'
```
# Select a value by row and column name

By label/location

```
>>>df.ix[2]
   Country      Brazil
   Capital      Brasília
   Population   207847528
```
# Select a row

```
>>>df.ix[:, 'Capital']              # Select a column
  0        Brussels
  1        New Delhi
  2        Brasília

>>>df.ix[1, 'Capital']
'New Delhi'
```

Boolean index
```
>>>s[~(s>1)]                        # There is no value greater than 1 in the
                                      sequence S
>>>s[(s<-1) | (s>2)]                # A value less than-1 or greater than 2 in the se-
                                      quence S
>>>df[df['Population']>1200000000]  # Use filters to adjust the data frame
```

Settings
```
>>>s['a']=6                         # Set the value of index a in sequence S to 6
```

(3) delete data
```
>>>s.drop(['a', 'c'])               # Delete the value of the sequence by index
                                      (axis=0)
>>>df.drop('Country', axis=1)       # Delete the column of the data frame by column
                                      name(axis=1)
```

(4) Sorting and ranking
```
>>>df.sort_index()                  # Sort by index
>>>df.sort_values(by='Country')     # Sort by the value of a column
>>>df.rank()                        # Data frame ranking
```

(5) Query sequence and data frame information

- **Basic Information**
```
>>>df.shape                         # (Row, column))
>>>df.index                         # Get the index
>>>df.columns                       # Get the column name
>>>df.info()                        # Get the basic information of the data frame
>>>df.count()                       # The number of non-Na values
```

- **Summary**
```
>>>df.sum()                         # Total
>>>df.cumsum()                      # cumulative
>>>df.min()/df.max()                # Minimum divided by maximum
>>>df.idxmin()/df.idxmax()          # Index minimum divided by index maximum
>>>df.describe()                    # Basic statistics
>>>df.mean()                        # average value
>>>df.median()                      # Median
```

- **Application function**

```
>>>f=lambda x: x*2              # Apply anonymous function lambda
>>>df.apply(f)                   # Application function
>>>df.applymap(f)                # Apply function to each cell
```

### A2.2.2.3 Input/Output

(1) Read/write CSV

```
>>>pd.read_csv('file.csv', header=None, nrows=5)
>>>df.to_csv('myDataFrame.csv')
```

(2) Read/write to Excel

```
>>>pd.read_excel('file.xlsx')
>>>pd.to_excel('dir/myDataFrame.xlsx', sheet_name='Sheet1')
```

**Read Excel with multiple tables**

```
>>>xlsx=pd.ExcelFile('file.xls')
>>>df=pd.read_excel(xlsx, 'Sheet1')
```

(3) Read and write SQL queries and database tables

```
>>>from sqlalchemy import create_engine
>>>engine=create_engine('sqlite:///:memory:')
>>>pd.read_sql("SELECT * FROM my_table;", engine)
>>>pd.read_sql_table('my_table', engine)
>>>pd.read_sql_query("SELECT * FROM my_table;", engine)
>>>pd.to_sql('myDf', engine)
```

### A2.2.3 SciPy

SciPy is a Python scientific computing core library based on Numpy.

### A2.2.3.1 Interact with Numpy

```
>>>import numpy as np
>>>a=np.array([1, 2, 3])
>>>b=np.array([(1+5j, 2j, 3j), (4j, 5j, 6j)])
>>>c=np.array([[(1.5, 2, 3), (4, 5, 6)], [(3, 2, 1), (4, 5, 6)]])
```

### A2.2.3.2 Manipulate shapes

```
>>>np.transpose(b)        # transpose of a matrix
>>>b.flatten()            # Flatten array
>>>np.hstack((b, c))      # Stack arrays horizontally in columns
                          # Stack arrays vertically in rows
>>>np.vstack((a, b))      # Divide the array horizontally at index 2
>>>np.hsplit(c, 2)
                          # Split the array vertically at index 2
>>>np.vpslit(d, 2)
```

### A2.2.3.3 Linear algebra

```
>>> from scipy import linalg, sparse
```

(1) Create matrix

```
>>> A = np.matrix (np.random.random((2, 2)))
>>> B = np.asmatrix(b)
>>> C = np.mat (np.random.random((10, 5)))
>>> D = np.mat([[3, 4], [5, 6]])
```

(2) Basic matrix routine

Inverse matrix

```
>>> A.I                              # Inversion matrix
>>> linalg.inv(A)                    # Inversion matrix
>>> A.T                              # matrix transpose
>>> A.H                              # conjugate transpose
>>> np.trace(A)                      # Calculate the sum of diagonal elements
```

Norm

```
>>> linalg.norm(A)                   # Frobenius Norm
>>> linalg.norm(A, 1)                # L1 norm (maximum column summary)
>>> linalg.norm(A, np.inf)           # Lnorm (maximum column summary)
```

ranking

```
>>> np.linalg.matrix_rank(C)         # Matrix ranking
```

Determinant

```
>>> linalg.det(A)                    # Determinant
```

Solve linear problems

```
>>> linalg.solve(A, b)               # Solving dense matrix
>>> E = np.mat(a).T                  # Solving dense matrix
>>> linalg.lstsq(D, E)               # Solving linear algebraic equations by least square method
```

Generalized inverse

```
>>> linalg.pinv(C)                   # Calculate the pseudo-inverse of the matrix (least square solver)
>>> linalg.pinv2(C)                  # Calculate the pseudo inverse (SVD) of matrix
```

(3) Create sparse matrix

```
>>> F = np.eye(3, k=1)               # Create 2X2 identity matrix
>>> G = np.mat(np.identity(2))       # Create 2X2 identity matrix
>>> C[C>0.5] = 0
>>> H = sparse.csr_matrix(C)         # Compress sparse row matrix
>>> I = sparse.csc_matrix(D)
>>> J = sparse.dok_matrix(A)         # Compress sparse row matrix
```

```
>>>E.todense()                      #DOK matrix
                                    #Convert sparse matrix to full matrix
>>>sparse.isspmatrix_csc(A)         #Unit sparse matrix
```
(4) Sparse matrix routine
Inverse matrix
```
>>>sparse.linalg.inv(I)             #Inversion matrix
```
Norm
```
>>>sparse.linalg.norm(I)            #Norm
```
Solve linear problems
```
>>>sparse.linalg.spsolve(H, I)      #Sparse solution sparse matrix
```
(5) Sparse matrix function
```
>>>sparse.linalg.expm(I)            #Sparse matrix index
```
(6) Invoke help
```
>>>help (scipy.linalg.diagsvd)
>>>np.info (np.matrix)
```
(7) Matrix function
Addition
```
>>>np.add(A, D)
```
Subtraction
```
>>>np.subtract(A, D)
```
Division
```
>>>np.divide(A, D)
```
Multiplication
```
>>>np.multiply(D, A)
>>>np.dot(A, D)                     #dot product
>>>np.vdot(A, D)                    #Vector dot product
>>>np.inner(A, D)                   #Inner product
>>>np.outer(A, D)                   #Outer product
>>>np.tensordot(A, D)               #Tensor dot product
>>>np.kron(A, D)                    #Kronecker multiplication
```

## A2.2.4 Scikit-learn

Scikit-learn is an open-source Python library, which realizes machine learning, preprocessing, cross-validation, and visualization algorithms through a unified interface. There are rich examples on the homepage of Scikit-learn. Here we just give a brief introduction. Examples:
```
>>>from sklearn import neighbors, datasets, preprocessing
>>>from sklearn.model_selection import train_test_split
>>>from sklearn.metrics import accuracy_score
```

```
>>> iris = datasets.load_iris()
>>> X, y = iris.data[:, :2], iris.target
>>> X_train, X_test, y_train, y_test = train_test_split(X, y, random_state=33)
>>> scaler = preprocessing.StandardScaler().fit(X_train)
>>> X_train = scaler.transform(X_train)
>>> X_test = scaler.transform(X_test)
>>> knn = neighbors.KNeighborsClassifier(n_neighbors=5)
>>> knn.fit(X_train, y_train)
>>> y_pred = knn.predict(X_test)
>>> accuracy_score(y_test, y_pred)
```

#### A2.2.4.1　Load data

The data processed by Scikit-learn are numbers stored as Numpy array or SciPy sparse matrix, and also support other data types such as Pandas data frame which can be converted into number array.

```
>>> import numpy as np
>>> X = np.random.random((10, 5))
>>> y = np.array(['M', 'M', 'F', 'F', 'M', 'F', 'M', 'M', 'F', 'F', 'F'])
>>> X[X < 0.7] = 0
```

#### A2.2.4.2　Generation of training and testing dataset

```
>>> from sklearn.model_selection import train_test_split
>>> X_train, X_test, y_train, y_test = train_test_split(X, y, testrandom_state=0)
```
# testing size is set to 25% of the total dataset.

#### A2.2.4.3　Create model

(1) Supervised learning evaluator

- **Linear regression**

```
>>> from sklearn.linear_model import LinearRegression
>>> lr = LinearRegression(normalize=True)
```

- **Support vector machine (SVM)**

```
>>> from sklearn.svm import SVC
>>> svc = SVC(kernel='linear')
```

- **Naive Bayes**

```
>>> from sklearn.naive_bayes import GaussianNB
>>> gnb = GaussianNB()
```

- **KNN**

```
>>> from sklearn import neighbors
>>> knn = neighbors.KNeighborsClassifier(n_neighbors=5)
```

(2) Unsupervised learning evaluator

- **Principal component analysis (PCA)**

\>\>\>from sklearn.decomposition import PCA
\>\>\>pca=PCA(n_components=0.95)

- **K Means**

\>\>\>from sklearn.cluster import KMeans
\>\>\>k_means=KMeans (n_clusters=3, random_state=0)

### A2.2.4.4 Model fitting

(1) Supervised learning

\>\>\>lr.fit(X, y)
\>\>\>knn.fit(X_train, y_train)
\>\>\>svc.fit(X_train, y_train)        # Fitting data and model

(2) Unsupervised learning

\>\>\>k_means.fit(X_train)              # Fitting data and model
\>\>\>pca_model=pca.fit_transform(X_train)   # Converted data to be merged

### A2.2.4.5 Prediction

(1) Supervised evaluator

\>\>\>y_pred=svc.predict(np.random.random((2, 5)))   # Forecast label
\>\>\>y_pred=lr.predict(X_test)          # Forecast label
\>\>\>y_pred=knn.predict_proba(X_test)   # Evaluate tag probability

(2) Unsupervised evaluator

\>\>\>y_pred=k_means.predict(X_test)     # Label in predictive clustering

### A2.2.4.6 Evaluate model performance

(1) Classification index

- **Accuracy rate**

\>\>\>knn.score(X_test, y_test)          # Evaluator scoring method
\>\>\>from sklearn.metrics import accuracy_score   # Index scoring function
\>\>\>accuracy_score(y_test, y_pred)

- **Classification prediction evaluation function**

\>\>\>from sklearn.metrics import classification_report   # Accuracy, recall rate, F1
\>\>\>print(classification_report(y_test, y_pred))        score and support rate

- **Confusion Matrix**

\>\>\>from sklearn.metrics import confusion_matrix
\>\>\>print(confusion_matrix(y_test, y_pred))

(2) Regression index

- **Mean absolute error**

\>\>\>from sklearn.metrics import mean_absolute_error\>\>\>y_true=[3, −0.5, 2]
\>\>\>mean_absolute_error(y_true, y_pred)

- **Mean square error**

>>>from sklearn. metrics import mean_squared_error

>>>mean_squared_error(y_test, y_pred)

- **$R^2$ score**

>>>from sklearn. metrics import r2_score

>>>r2_score(y_true, y_pred)

(3) Cluster index

- **Adjust rand coefficient**

>>>from sklearn. metrics import adjusted_rand_score

>>>adjusted_rand_score(y_true, y_pred)

- **homogeneity**

>>>from sklearn. metrics import homogeneity_score

>>>homogeneity_score(y_true, y_pred)

- **V-measure**

>>>from sklearn. metrics import v_measure_score

>>>metrics. v_measure_score(y_true, y_pred)

### A2.2.5　Matplotlib

Matplotlib is a two-dimensional drawing library of Python, which is used to generate all kinds of graphics that meet publishing quality or cross-platform interactive environment.

#### A2.2.5.1　Prepare data

(1) One dimensional data

>>>import numpy as np

>>>x=np. linspace(0, 10, 100)

>>>y=np. cos(x)

>>>z=np. sin(x)

(2) Two-dimensional data or pictures

>>>data=2 * np. random. random((10, 10))

>>>data2=3 * np. random. random((10, 10))

>>>Y, X=np. mgrid[-3: 3: 100j, -3: 3: 100j]

>>>U=-1-X**2+Y

>>>V=1+X-Y**2

>>>from matplotlib. cbook import get_sample_data

>>>img=np. load(get_sample_data('axes_grid/bivariate_normal. npy'))

#### A2.2.5.2　Draw graphics

>>>import matplotlib. pyplot as plt

(1) Canvas

>>>fig=plt. figure()

>>>fig2=plt. figure(figsize=plt. figaspect(2.0))

(2) axis

Graphics are drawn with axes as the core. In most cases, subgraphs can meet the needs. A subgraph is the coordinate axis of a grid system.

```
>>>fig.add_axes()
>>>ax1=fig.add_subplot(221)
>>>ax3=fig.add_subplot(212)          # 212 means the second sub-
>>>fig3, axes=plt.subplots(nrows=2, ncols=2)   graph occupying the second row
>>>fig4, axes2=plt.subplots(ncols=3)           of the first column
```

### A2.2.5.3 Drawing routine

(1) One dimensional data

```
>>>fig, ax=plt.subplots()            # Mark connection points
                                       with lines or marks
>>>lines=ax.plot(x, y)               # Draw a vertical rectangle of
>>>ax.scatter(x, y)                    equal width
>>>axes[0, 0].bar([1, 2, 3], [3, 4, 5])   # Draw a horizontal rectangle
                                            with equal height
>>>axes[1, 0].barh([0.5, 1, 2.5], [0, 1, 2])   # Draw a horizontal line paral-
                                                 lel to the axis
>>>axes[1, 1].axhline(0.45)          # Draw a horizontal line per-
>>>axes[0, 1].axvline(0.65)            pendicular to the axis
                                       # Draw a filled polygon
>>>ax.fill(x, y, color='blue')
                                       # Fill in between y value and 0
>>>ax.fill_between(x, y, color='yellow')
```

(2) Two-dimensional data or pictures

```
>>>fig, ax=plt.subplots()
>>>im=ax.imshow(img,
cmap='gist_earth', interpolation='nearest',   # Color table or RGB array
vmin=-2, vmax=2)                              # Two-dimensional array pseu-
>>>axes2[0].pcolor(data2)                       do-color diagram
                                                # Pseudo-color map of two-di-
>>>axes2[0].pcolormesh(data)                    mensional array contour line

>>>CS=plt.contour(Y, X, U)           # contour map
>>>axes2[2].contourf(data1)          # Contour map label
>>>axes2[2]=ax.clabel(CS)
```

(3) Vector field

```
>>>axes[0, 1].arrow(0, 0, 0.5, 0.5)   # Add arrows to axes
```

```
>>>axes[1,1].quiver(y,z)              # Two dimensional arrow
>>>axes[0,1].streamplot(X,Y,U,V)      # Two dimensional arrow
```
(4) Data distribution
```
>>>ax1.hist(y)                         # Histogram
>>>ax3.boxplot(y)                      # Box-plot
>>>ax3.violinplot(z)                   # Violin pictures
```

### A2.2.5.4 Custom graphics

(1) Colors, color bars and color tables
```
>>>plt.plot(x, x, x, x**2, x, x**3)
>>>ax.plot(x, y, alpha=0.4)
>>>ax.plot(x, y, c='k')
>>>fig.colorbar(im, orientation='horizontal')
>>>im=ax.imshow(img, cmap='seismic')
```
(2) Sign/marker
```
>>>fig, ax=plt.subplots()
>>>ax.scatter(x, y, marker=".")
>>>ax.plot(x, y, marker="o")
```
(3) linetype
```
>>>plt.plot(x, y, linewidth=4.0)
>>>plt.plot(x, y, ls='solid')
>>>plt.plot(x, y, ls='——')
>>>plt.plot(x, y, '——', x**2, y**2, '-.')
>>>plt.setp(lines, color='r', linewidth=4.0)
```
(4) Text and annotations
```
>>>ax.text(1, -2.1, 'Example Graph', style='italic')
>>>ax.annotate("Sine", xy=(8, 0), xycoords='data', xytext=(10.5, 0), textcoords='data', arrowprops=dict(arrowstyle="->", connectionstyle="arc3"),)
```
(5) Mathematical symbol
```
>>>plt.title(r'$ sigma_i=15 $', fontsize=20)
```
(6) Size restrictions, legends and layouts

- **Size limitation and automatic adjustment**
```
>>>ax.margins(x=0.0, y=0.1)            # Add inner margins
>>>ax.axis('equal')                     # Set the graphic aspect ratio to 1
>>>ax.set(xlim=[0,10.5], ylim=[-1.5,1.5]) # Set limits for x and y axes
>>>ax.set_xlim(0, 10.5)                # Set limits for the x
```
- **Legend**
```
>>>ax.set(title='An Example Axes',     # Set the label of title and x and y axes
          ylabel='Y-Axis',
          xlabel='X-Axis')
```

# Appendix

```
>>>ax.legend(loc='best')         # Automatically select the best legend position
```

- **Sign**

```
>>>ax.xaxis.set(ticks=range(1, 5),          # Manually set the x-axis
                ticklabels=[3,100,-12,"foo"])    scale
>>>ax.tick_params(axis='y',
                  direction='inout',        # Set the y-axis length and
                  length=10)                direction
```

- **Subgraph spacing**

```
>>>fig3.subplots_adjust(wspace=0.5,
                        hspace=0.3,
                        left=0.125,
                        right=0.9,
                        top=0.9,            # Adjust the distance between
                        bottom=0.1)         subgraphs
>>>fig.tight_layout()               # Set the subgraph layout of
                                    canvas
```

- **Axis edge**

```
>>>ax1.spines['top'].set_visible(False)       # Hide the top coordinate
                                              axis
>>>ax1.spines['bottom'].set_position(('outward', 10))  # Set the position of the
                                                       bottom edge to outward
```

### A2.2.5.5 Save and show plots

(1) Save canvas

```
>>>plt.savefig('foo.png')
```

**Save transparent canvas**

```
>>>plt.savefig('foo.png', transparent=True)
```

(2) Show graphics

```
>>>plt.show()
```

(3) Closing and clearing

```
>>>plt.cla()          # Clear axes
>>>plt.clf()          # Clear canvas
>>>plt.close()        # Close window
```

## A3  Linear Algebra

Linear algebra can provide a useful way of compactly representing and operating on sets of linear equations. There are many advantages of using this compact format such as space savings. In the following, we will present some basic operations and properties. Before moving forward to these introductions, we first give the notations of matrix and vectors.

- $A$: a matrix with $m$ rows and $n$ columns. Its entry is denoted by $a_{ij}$ representing the in the $i$th row and $j$th column.
- $x$: a vector with $n$ entries. By convention, an $n$-dimensional vector is often thought of as a matrix with $n$ rows and 1 column, known as a column vector. To represent a row vector, we can write $x^T$. The $i$th element of a vector $x$ is denoted as $x_i$.

### A3.1  The Identity Matrix and Diagonal Matrices

The $n \times n$ identity matrix, denoted $I \in \mathbb{R}^{n \times n}$, is a square matrix with ones on the diagonal and zeros otherwise, i.e.:

$$I_{ij} = \begin{cases} 1 & i=j \\ 0 & i \neq j \end{cases}$$

It has the property that for all $A \in \mathbb{R}^{m \times n}$, $AI = A = IA$

A diagonal matrix is a matrix where all non-diagonal elements are 0. This is typically denoted $D = diag(d_1, d_2, \ldots, d_n)$, with

$$D_{ij} = \begin{cases} d_i & i=j \\ 0 & i \neq j \end{cases}$$

Obviously, $I = diag(1, 1, \ldots, 1)$, so we can say that the identity matrix is a special case of a diagonal matrix.

### A3.2  The Transpose

Given a matrix $A \in \mathbb{R}^{m \times n}$, its transpose, written as $A^T \in \mathbb{R}^{n \times m}$, is the $n \times m$ matrix whose entries are given by: $(A^T)_{ij} = A_{ji}$. The transpose of a matrix has the following verified properties:

- $(A^T)^T = A$
- $(AB)^T = B^T A^T$
- $(A+B)^T = A^T + B^T$

### A3.3  Symmetric Matrices

A square matrix $A \in \mathbb{R}^{n \times n}$ is symmetric if $A = A^T$. It is anti-symmetric if $A = -A^T$. It is easy to know that for any matrix $A \in \mathbb{R}^{n \times n}$ the matrix $A + A^T$ is symmetric and the matrix $A - A^T$ is anti-symmetric. Specifically, any square matrix $A \in \mathbb{R}^{n \times n}$ can be written as a sum of a symmetric matrix and an anti-symmetric matrix:

$$A = \frac{1}{2}(A + A^T) + \frac{1}{2}(A - A^T)$$

where the first matrix on the right side is symmetric, while the second term is anti-symmetric.

### A3.4  The Trace

The trace of a square matrix $A \in \mathbb{R}^{n \times n}$, usually denoted $tr(A)$ (or just $trA$), is the sum of diagonal elements in the matrix, that is:

$$trA = \sum_{i=1}^{n} A_{ii}$$

The trace has the following properties:
- For $A \in \mathbb{R}^{n \times n}$, $trA = trA^T$.
- For $A, B \in \mathbb{R}^{n \times n}$, $tr(A+B) = trA + trB$.
- For $A \in \mathbb{R}^{n \times n}$, $t \in \mathbb{R}$, $tr(tA) = ttrA$.
- For $A, B$ such that $AB$ is square, $trAB = trBA$, and so for the product of more matrices.

### A3.5 Norms

A norm of a vector $\|x\|$ is informally a measure of the "length" of the vector. The $\ell_2$ norm is

$$\|x\|_2 = \sqrt{\sum_{i=1}^{n} x_i^2}$$

$\ell_1$ norm:

$$\|x\|_1 = \sum_{i=1}^{n} |x_i|$$

and the $\ell_\infty$ norm:

$$\|x\|_\infty = \max_i |x_i|$$

Note all the three norms above are concrete examples of the $\ell_p$ norm, which is defined as:

$$\|x\|_p = \left(\sum_{i=1}^{n} |x_i|^p\right)^{1/p}$$

### A3.6 Linear Independence and Rank

A set of vectors $x_1, x_2, \ldots, x_n \in \mathbb{R}$ is said to be linearly independent if no vector can be represented as a linear combination of the other remaining vectors. Conversely, if one vector belonging to the set can be represented as a linear combination of the rest vectors, then the vectors are said to be linearly dependent: $x_n = \sum_{i=1}^{n-1} x_i$.

The column rank of a matrix $A \in \mathbb{R}^{m \times n}$ is the size of the largest subset of columns of $A$ that constitute a linearly independent set. Similarly, the row rank is the largest number of rows of $A$ that constitute a linearly independent set. If $rank(A) = \min(m, n)$, then $A$ is said to be full rank.

### A3.7 The Inverse of a Square Matrix

The inverse of a square matrix $A \in \mathbb{R}^{n \times n}$ is denoted as $A^{-1}$, and is the unique matrix

such that.
$$A^{-1}A = I = AA^{-1}$$

The following are properties of the inverse; all assume that $A$, $B \in \mathbb{R}^{n \times n}$ are non-singular:
$$(A^{-1})^{-1} = A$$
$$(AB)^{-1} = B^{-1}A^{-1}$$
$$(A^{-1})^{\mathrm{T}} = (A^{\mathrm{T}})^{-1}$$

As an example of how the inverse is used, consider the linear system of equations, $Ax = b$ where $A \in \mathbb{R}^{n \times n}$, and $x$, $b \in \mathbb{R}$. If $A$ is nonsingular (i.e., invertible), then $x = A^{-1}b$.

### A3.8 Orthogonal Matrices

Two vectors $x$, $y \in \mathbb{R}^n$ are orthogonal if $x^{\mathrm{T}}y = 0$. A vector $x \in \mathbb{R}^n$ is normalized if $\|x\|_2 = 1$. A square matrix $U \in \mathbb{R}^{n \times n}$ is orthogonal (note the different meanings when talking about vectors versus matrices) if all its columns are orthogonal to each other and are normalized (the columns are then referred to as being orthonormal).

It follows immediately from the definition of orthogonality and normality that
$$U^{\mathrm{T}}U = I = UU^{\mathrm{T}}$$

### A3.9 The Determinant

The determinant of a square matrix $A \in \mathbb{R}^{n \times n}$ is a function det: $\mathbb{R}^{n \times n} \to \mathbb{R}$, and is denoted $|A|$ or det $A$. Some important properties include:
- For identity, the determinant is 1: $|I| = 1$.
- For $A \in \mathbb{R}^{n \times n}$, $|A| = |A^{\mathrm{T}}|$.
- For $A$, $B \in \mathbb{R}^{n \times n}$, $|AB| = |A||B|$.
- For $A \in \mathbb{R}^{n \times n}$, $|A| = 0$ if and only if $A$ is singular (i.e., non-invertible).
- For $A \in \mathbb{R}^{n \times n}$ and $A$ non-singular, $|A^{-1}| = 1/|A|$.

### A3.10 Quadratic Forms

Given a square matrix $A \in \mathbb{R}^{n \times n}$ and a vector $x \in \mathbb{R}^n$, the scalar value $x^{\mathrm{T}}Ax$ is called a quadratic form:
$$x^{\mathrm{T}}Ax = \sum_{i=1}^{n} x_i(Ax)_i = \sum_{i=1}^{n} x_i \left( \sum_{j=1}^{n} A_{ij}x_j \right) = \sum_{i=1}^{n}\sum_{j=1}^{n} A_{ij}x_ix_j$$

The matrix $A$ here appearing in the quadratic form is often symmetric.

### A3.11 Eigenvalues and Eigenvectors

Given a square matrix $A \in \mathbb{R}^{n \times n}$, we say that $\lambda \in \mathbb{C}$ is an eigenvalue of $A$ and $x \in \mathbb{C}^n$ is the corresponding eigenvector if
$$Ax = \lambda x, \quad x \neq 0$$

We can rewrite the equation above to:
$$(\lambda I - A)x = 0, x \neq 0$$

Since $x$ does not equal to 0, the make the equation set up the $(\lambda I - A)$ should be singular, that is $|(\lambda I - A)| = 0$.

### A3.12 Matrix Calculus

#### A3.12.1 The Gradient

Suppose that $f: \mathbb{R}^{m \times n} \to \mathbb{R}$ is a function that takes as input a matrix $A$ of size $m \times n$ and returns a real value. Then the gradient of $f$ (with respect to $A \in \mathbb{R}^{m \times n}$) is the matrix of partial derivatives, defined as:

$$\nabla_A f(A) \in \mathbb{R}^{m \times n} = \begin{bmatrix} \frac{\partial f(A)}{\partial A_{11}} & \frac{\partial f(A)}{\partial A_{12}} & \cdots & \frac{\partial f(A)}{\partial A_{1n}} \\ \frac{\partial f(A)}{\partial A_{21}} & \frac{\partial f(A)}{\partial A_{22}} & \cdots & \frac{\partial f(A)}{\partial A_{2n}} \\ \vdots & \vdots & \ddots & \vdots \\ \frac{\partial f(A)}{\partial A_{m1}} & \frac{\partial f(A)}{\partial A_{m2}} & \cdots & \frac{\partial f(A)}{\partial A_{mn}} \end{bmatrix}$$

i.e., an $m \times n$ matrix with

$$(\nabla_A f(A))_{ij} = \frac{\partial f(A)}{\partial A_{ij}}$$

when $A$ is a vector $x \in \mathbb{R}^n$, the gradient becomes:

$$\nabla_x f(x) = \begin{bmatrix} \frac{\partial f(x)}{\partial x_1} \\ \frac{\partial f(x)}{\partial x_2} \\ \vdots \\ \frac{\partial f(x)}{\partial x_n} \end{bmatrix}$$

#### A3.12.2 The Hessian

Suppose that $f: \mathbb{R}^n \to \mathbb{R}$ is a function that takes a vector in $\mathbb{R}^n$ and returns a real number. Then the Hessian matrix with respect to $x$, written $\nabla_x^2 f(Ax)$ or simply as $H$ is the $n \times n$ matrix of partial derivatives,

$$\nabla_x^2 f(x) \in \mathbb{R}^{n \times n} = \begin{bmatrix} \frac{\partial^2 f(x)}{\partial x_1^2} & \frac{\partial^2 f(x)}{\partial x_1 \partial x_2} & \cdots & \frac{\partial^2 f(x)}{\partial x_1 \partial x_n} \\ \frac{\partial^2 f(x)}{\partial x_2 \partial x_1} & \frac{\partial^2 f(x)}{\partial x_2^2} & \cdots & \frac{\partial^2 f(x)}{\partial x_2 \partial x_n} \\ \vdots & \vdots & \ddots & \vdots \\ \frac{\partial^2 f(x)}{\partial x_n \partial x_1} & \frac{\partial^2 f(x)}{\partial x_n \partial x_2} & \cdots & \frac{\partial^2 f(x)}{\partial x_n^2} \end{bmatrix}$$

for each element of $\nabla_x^2 f(x) \in \mathbb{R}^{n \times n}$:

$$(\nabla_x^2 f(x))_{ij} = \frac{\partial^2 f(x)}{\partial x_i \partial x_j}$$

The Hessian is always symmetric for

$$\frac{\partial^2 f(\boldsymbol{x})}{\partial \boldsymbol{x}_i \partial \boldsymbol{x}_j} = \frac{\partial^2 f(\boldsymbol{x})}{\partial \boldsymbol{x}_j \partial \boldsymbol{x}_i}$$

For the $i$th entry of the gradient $(\nabla_{\boldsymbol{x}} f(\boldsymbol{x}))_i$, the gradient of it with respect to $\boldsymbol{x}$ is

$$\nabla_{\boldsymbol{x}} \frac{\partial f(\boldsymbol{x})}{\partial \boldsymbol{x}_i} = \begin{bmatrix} \dfrac{\partial^2 f(\boldsymbol{x})}{\partial \boldsymbol{x}_i \partial \boldsymbol{x}_1} \\ \dfrac{\partial^2 f(\boldsymbol{x})}{\partial \boldsymbol{x}_i \partial \boldsymbol{x}_2} \\ \vdots \\ \dfrac{\partial^2 f(\boldsymbol{x})}{\partial \boldsymbol{x}_i \partial \boldsymbol{x}_n} \end{bmatrix}$$

# A4  Probability

## A4.1  Random Variable and Probability Distribution

A random variable is a function that assigns numbers to the results of an experiment. Each possible outcome of the experiment occurs with a certain probability. This outcome variable is a random variable because it is uncertain what value it will take. Probabilities are associated with the outcomes to quantify this uncertainty.

A random variable is called discrete if the set of all possible outcomes $x_1, x_2 \ldots$ is finite or countable. For a discrete random variable, $X$, a probability density function is defined to be the function $f(x_i)$. For any value $x_i$ that $X$ can take, $f$ gives the probability when the random variable $X$ is equal to $x_i$. Mathematically,

$$P(X = x_i) = f(x_i) \quad i = 1, 2, \ldots \tag{A4.1a}$$

$$f(x_i) \geqslant 0, \quad \sum_i f(x_i) = 1 \tag{A4.1b}$$

When $X$ can take any values in at least one interval on the real number, it is called a continuous random variable. Since the possible values of $X$ are uncountable, the probability associated with any particular point is zero. Unlike the situation for discrete random variables, the density function of a continuous random variable $X$ will not give the probability that $X$ takes the value $x_i$. Instead, the probability density function of a continuous random variable is defined as $f(x) \geqslant 0$ and

$$P(a < X \leqslant b) = \int_a^b f(x) \mathrm{d}x; \ a \leqslant b \tag{A4.1c}$$

This is the area under $f(x)$ in the range from $a$ to $b$, which satisfy

$$\int_{-\infty}^{+\infty} f(x) \mathrm{d}x = 1 \tag{A4.1d}$$

## A4.2  Cumulative Distribution Function

A function closely related to the probability density function of a discrete random variable is the corresponding cumulative distribution function.

$$F(x)=P(X\leqslant x)=\sum\nolimits_{X\leqslant x}f(X) \tag{A4.2b}$$

The cumulative distribution function for a continuous random variable $X$ is given by

$$F(x)=P(X\leqslant x)=\int_{-\infty}^{x}f(t)\,dt \tag{A4.2b}$$

where $f(t)$ is the probability density function. In both the continuous and the discrete case, $F(x)$ must satisfy the following properties:
- $0\leqslant F(x)\leqslant 1$.
- if $x_2\geqslant x_1$ then $F(x_2)\geqslant F(x_1)$.
- $F(+\infty)=1$ and $F(-\infty)=0$.

### A4.3 Expectations of Random Variables

The expected value of a random variable $X$, denoted by $E(X)$, is a weighted average of the values taken by the random variable $X$, where the weights are the respective probabilities. Let us consider the discrete random variable $X$ with outcomes $x_1, x_2, \ldots, x_n$ and corresponding probabilities $f(x_i)$. Then, the expression of $E(X)$ is

$$E(X)=\mu=\sum_{i=1}^{n}x_i f(X=x_i) \tag{A4.3a}$$

which defines the expected value of the discrete random variable. For a continuous random variable $X$ with density $f(x)$, the expected value becomes:

$$E(X)=\mu=\int_{-\infty}^{+\infty}xf(x)\,dx \tag{A4.3b}$$

### A4.4 Joint Distribution Function

The probability function defined over a pair of random variables is called the joint probability distribution. Consider two random variables $X$ and $Y$, the joint probability distribution function of $X$ and $Y$ is defined as the probability that $X$ is equal to $x_i$ at the same time that $Y$ is equal to $y_i$.

$$P(\{X=x_i\}\cap\{Y=y_j\})=P(X=x_i,Y=y_j)=f(x_i,y_j)\quad i,j=1,2,\ldots \tag{A4.4a}$$

If $X$ and $Y$ are continuous random variables, then the bivariate probability density function is:

$$P(a<X\leqslant b;c<Y\leqslant d)=\int_{c}^{a}\int_{a}^{b}f(x,y)\,dxdy \tag{A4.4b}$$

The counterparts of the requirements for a probability density function are:

$$\sum_i\sum_j f(x_i,y_j)=1 \tag{A4.5c}$$

$$\int_{-\infty}^{+\infty}\int_{-\infty}^{+\infty}f(x,y)\,dxdy=1 \tag{A4.5d}$$

The cumulative joint distribution function in the case that both $X$ and $Y$ are discrete random variables is

$$F(x,y)=P(X\leqslant x,Y\leqslant y)=\sum_{X\leqslant x}\sum_{Y\leqslant y}f(X,Y) \tag{A4.4e}$$

and if both $X$ and $Y$ are continuous random variables then

$$F(x,y)=P(X\leqslant x,Y\leqslant y)=\int_{-\infty}^{x}\int_{-\infty}^{y}f(t,v)\mathrm{d}t\mathrm{d}v \tag{A4.4f}$$

### A4.5 Marginal Distribution Function

The marginal distribution, $f(x)$, of a discrete random variable $X$ provides the probability that $X$ is equal to $x$ in the joint probability $f(X,Y)$, without considering the variable $Y$, thus, if we want to obtain the marginal distribution of $X$ from the joint density, it is necessary to sum out the other variable $Y$. The marginal distribution for the random variable $Y$, $f(y)$ is defined analogously.

$$P(X=x)=f(x)=\sum_{Y}f(x,Y) \tag{A4.5a}$$

$$P(Y=y)=f(y)=\sum_{X}f(X,y) \tag{A4.5b}$$

The resulting marginal distributions are one-dimensional. Similarly, we obtain the marginal densities for a pair of continuous random variables $X$ and $Y$:

$$f(x)=\int_{-\infty}^{+\infty}f(x,y)\mathrm{d}y \tag{A4.5c}$$

$$f(y)=\int_{-\infty}^{+\infty}f(x,y)\mathrm{d}x \tag{A4.5d}$$

### A4.6 Conditional Probability Distribution Function

In the setting of a joint bivariate distribution $f(X,Y)$, the probability distribution of $Y$ given $X=x$ is called the conditional probability distribution function of $Y$ given $X$, $F_{Y|X=x}(y)$. In the discrete case, it is defined as:

$$F_{Y|X=x}(y)=P(Y\leqslant y|X=x)=\sum_{Y\leqslant y}\frac{f(x,Y)}{f(x)}=\sum_{Y\leqslant y}f(Y|x) \tag{A4.6a}$$

For the continuous case, $F_{Y|X=x}(y)$ is defined as:

$$F_{Y|X=x}(y)=P(Y\leqslant y|X=x)=\int_{-\infty}^{y}f(y|x)\mathrm{d}y=\int_{-\infty}^{y}\frac{f(x,y)}{f(x)}\mathrm{d}y \tag{A4.6b}$$

where $f(Y|x)$ and $f(y|x)$ are the conditional probability density functions and $f(x)>0$.

### A4.7 Conditional Expectation

For a pair of random variables $(X,Y)$ with conditional probability density function, namely $f(y|x)$, the conditional expectation is defined as the expected value of the conditional distribution, i.e.

$$E(Y|X=x)=\begin{cases}\sum_{j=1}^{n}y_{j}f(Y=y_{j}|X=x) & \text{if } Y \text{ is discrete.}\\ \int_{-\infty}^{+\infty}yf(y|x)\mathrm{d}y & \text{if } Y \text{ is continuous.}\end{cases} \tag{A4.7}$$

### A4.8 The Regression Function

Given a pair of random variables $(X,Y)$ with a range of possible values, a regression function is that relates different values of $X$, say $x_1, x_2, \ldots, x_n$, and their corresponding values in terms of the conditional expectation $E(Y|X=x_1), \ldots, E(Y|X=x_n)$. The regression function describes the dependence of a quantity $Y$ on the quantity $X$, assuming there is a one-directional dependence. The variable $X$ is referred to as regressor, explanatory variable, or independent variable, $Y$ is referred to as regress and explained variable, or dependent variable.

## A5  Mathematical Optimization Basics

### A5.1  Optimization Problems

An optimization problem (5.1) consists of finding a vector $\boldsymbol{x}=(x_1, x_2, \ldots, x_n)$ that minimizes an objective function $f: \boldsymbol{X} \subset \mathbb{R}^n \to \mathbb{R}$, searching only among a set of solutions that satisfy given constraints:

$$\min_{\boldsymbol{x}} f(\boldsymbol{x}), \quad \text{subject to:} \tag{A5.1a}$$

$$g_i(\boldsymbol{x}) \leqslant 0, \quad i=1,2,\ldots,m \tag{A5.1b}$$

$$h_j(\boldsymbol{x}) = 0, \quad j=1,2,\ldots,p \tag{A5.1c}$$

$$\boldsymbol{x} \in S \tag{A5.1d}$$

where $x_1, x_2, \ldots, x_n$ are the problem decision variables. Function $f$ is called the problem objective function, (A5.1b) are the *inequality constraints*, (A5.1c) the *equality constraints*, and (A5.1d) the *set constraints*.

A solution $\boldsymbol{x}$ that satisfies *all* problem constraints (A5.1b) ~ (A5.1d) is referred to as a feasible solution. The set of all feasible solutions to a problem is called the *feasible set*. All the functions $f$, $g_i$ and $h_j$ in the objective and constraints should be well defined in the feasible set. When the problem has no constraints, we refer to it as an *unconstrained* optimization problem, and every vector in $\mathbb{R}^n$ is feasible.

Given an optimization problem $\min_{\boldsymbol{x} \in \boldsymbol{X}} f(\boldsymbol{x})$, a vector $\boldsymbol{x}^* \in \boldsymbol{X}$ is said to be a *global* optimum, when $f(\boldsymbol{x}^*) \leqslant f(\boldsymbol{x})$ for all $\boldsymbol{x} \in \boldsymbol{X}$. $\boldsymbol{x}^* \in \boldsymbol{X}$ is said to be a local optimum when $f(\boldsymbol{x}^*) \leqslant f(\boldsymbol{x})$, for every $\boldsymbol{x} \in \boldsymbol{X} \cap \boldsymbol{B}(\boldsymbol{x}^*, \boldsymbol{\varepsilon})$, where $\boldsymbol{B}(\boldsymbol{x}^*, \boldsymbol{\varepsilon})$ is a ball centered in $\boldsymbol{x}^*$ and with a radius $\boldsymbol{\varepsilon} > 0$, as small as it is given. In other words, $f$ reaches a minimum in $\boldsymbol{x}^*$, compared to the feasible points which are in its proximity. Global and local minimums are labeled as strict when the relation $f(\boldsymbol{x}^*) \leqslant f(\boldsymbol{x})$ holds strictly when $\boldsymbol{x} \neq \boldsymbol{x}^*$.

**Theorem 5.1** (Weierstrass) If $f$ is a continuous function, and $X$ is a non-empty compact set, then the problem $\min_{\boldsymbol{x} \in \boldsymbol{X}} f(\boldsymbol{x})$ has at least a global optimum.

### A5.2  A Classification of Optimization Problems

Optimization problems admit multiple classifications. We follow a classical approach next, and briefly review *linear programs*, *convex programs*, *nonlinear programs*, and *integer programs*.

### A5.2.1 Linear Programming

Linear optimization problems have the form below:

$$\min_{x} c_1 x_1 + c_2 x_2 + \dots + c_n x_n, \quad \text{subject to:}$$
$$a_{i1} x_1 + a_{i2} x_2 + \dots + a_{in} x_n \leqslant b_i, \quad \forall_i = 1, 2, \dots, m$$
$$a'_{j1} x_1 + a'_{j2} x_2 + \dots + a'_{jn} x_n \leqslant b'_j, \quad \forall_j = 1, 2, \dots, p$$

where all functions $f$, $g_i$, $h_j$ are linear①, this can be expressed in a compact matricial form:

$$\min_{x} \boldsymbol{c}^{\mathrm{T}}, \quad \text{subject to:} \ \boldsymbol{A}\boldsymbol{x} \leqslant \boldsymbol{b}, \ \boldsymbol{A}'\boldsymbol{x} \leqslant \boldsymbol{b}'$$

where matrices $\boldsymbol{A} \in \mathbb{R}^{m \times n}$, $\boldsymbol{A}' \in \mathbb{R}^{p \times n}$, and vectors $\boldsymbol{c} \in \mathbb{R}^n$, $\boldsymbol{b} \in \mathbb{R}^m$, $\boldsymbol{b}' \in \mathbb{R}^p$, are input constants.

The points satisfying a linear *inequality* constraint are called a *semispace*. The points satisfying linear *equality* constraints are called a *hyperplane*. The set of feasible points in a linear *program* is *a polyhedron*. A semi-space $\{\boldsymbol{x} \in \mathbb{R}^n: \boldsymbol{a}^{\mathrm{T}}\boldsymbol{x} \leqslant b\}$ is a convex set since it is the sub-level set of the convex function $f(\boldsymbol{x}) = \boldsymbol{a}^{\mathrm{T}}\boldsymbol{x}$. This applies also if we reverse the *inequality* sign, sinc the function $-\boldsymbol{a}^{\mathrm{T}}\boldsymbol{x}$ is also convex, and thus has convex sub-level sets. Then, a *hyperplane* $\{\boldsymbol{x} \in \mathbb{R}^n: \boldsymbol{a}^{\mathrm{T}}\boldsymbol{x} = b\}$ is a convex set since it can be expressed as the intersection of two semispaces: $\{\boldsymbol{x} \in \mathbb{R}^n: \boldsymbol{a}^{\mathrm{T}}\boldsymbol{x} \leqslant b, \boldsymbol{a}^{\mathrm{T}}\boldsymbol{x} \geqslant b\}$. Finally, the *polyhedron* is the intersection of semispaces and hyperplanes, and thus is also convex.

**Theorem 5.2.1** Let $P$ be a linear programming problem, and $\boldsymbol{X}$ is a feasible set. Then:
- If the problem has optimum solutions, at least one of them is a vertex of $\boldsymbol{X}$.
- If $\boldsymbol{x}^1$, $\boldsymbol{x}^2$ are two optimum solutions, all the points in the segment between them are also optimum.

**Proposition 5.2.1** For a linear programming problem:

$$\{\min_{x} \boldsymbol{c}^{\mathrm{T}}\boldsymbol{x}, \text{subject to:} \ \boldsymbol{A}_1 \boldsymbol{x} \leqslant \boldsymbol{b}_1, \boldsymbol{A}_2 \boldsymbol{x} \leqslant \boldsymbol{b}_2\}$$

where $\boldsymbol{A}_1 \in \mathbb{R}^{m \times n}$, $\boldsymbol{A}_2 \in \mathbb{R}^{p \times n}$. If the problem has optimum solutions, then at least one of them has at most $m + p$ non-zero coordinates.

### A5.2.2 Convex Programs

A program $\min_{x \in X} f(x)$ or $\min_{x \in X} -f(x)$, is a convex program when $f$ is a convex function, and $X$ is a convex set. A sufficient condition for a problem of the form (A5.1) to be convex is:
- The objective function $f$ is convex.
- All functions in *inequality* constraints $g_i(\boldsymbol{x}) \leqslant 0$, are convex.
- Allfunctions in *equality* constraints $h_j(\boldsymbol{x}) = 0$ are affine (linear plus a constant).
- The $S$ set is a convex set.

---

① A function $f: \mathbb{R}^n \to \mathbb{R}$ is linear if it can be expressed as $f(\boldsymbol{x}) = \boldsymbol{c}^{\mathrm{T}}\boldsymbol{x} = c_1 x_1 + c_2 x_2 + \dots + c_n x_n$, where $\boldsymbol{c} = (c_1, c_2, \dots, c_n)$ is any constant vector.

Some of the important properties are enumerated:

**Proposition 5.2.2.1** If $x$ is a local optimum of a convex problem, then $x$ is also a global optimum.

**Proposition 5.2.2.2** If $x^1$ and $x^2$ are optimum solutions to a convex problem, all the points in the segment between them are also optimal.

**Proposition 5.2.2.3** If the objective function $f$ of a convex minimization program is strictly convex, then if a solution exists, it is unique.

### A5.2.3  Nonlinear Programming

Nonlinear programming(NLP) is the process of solving an optimization problem where some of the constraints or the objective function $f$ are nonlinear. It is defined in a similar way with an optimization problem:

$$\min_x f(x), \quad \text{subject to:} \tag{A5.2a}$$

$$g_i(x) \leqslant 0, i=1,2,\dots,m \tag{A5.2b}$$

$$h_j(x) = 0, i=1,2,\dots,p \tag{A5.2c}$$

### A5.2.4  Integer Programs

Optimization problems of the form:

$$\min_x f(x), \text{subject to:} \tag{A5.3a}$$

$$g_i(x) \leqslant 0, \ i=1,2,\dots,m \tag{A5.3b}$$

$$h_j(x) \leqslant 0, j=1,2,\dots,p \tag{A5.3c}$$

$$x_k \in \mathbb{Z}, \forall k \in I \tag{A5.3d}$$

are called *integer programs* when all of the decision variables are constrained to be integers, and mixed-integer programs when only some of them are. In (A5.3), $I$ represents the set of integer constrained variables. When a decision variable is restricted to take just the values zero or one, it is a binary variable. When all the decision variables are binary, the problem is referred to as a *binary programming* problem.

## A5.3  Duality and dual function

Consider an optimization problem:

$$\min_x f(x), \quad \text{subject to:} \tag{A5.4a}$$

$$\pi_i : g_i(x) \leqslant 0, \ i=1,2,\dots,m \tag{A5.4b}$$

$$\lambda_j : h_j(x) = 0, \ j=1,2,\dots,p \tag{A5.4c}$$

$$x \in S \tag{A5.4d}$$

where $f$, $g_i$, $h_j$, are continuous functions (not necessarily convex) correctly defined in the problem feasibility set and $S$ is an arbitrary set. We name problem (A5.4) the primal problem, refer to $x$ as the primal variables.

We define the *Lagrangian function* of problem (A5.4) as a function $L: S \times \mathbb{R}_+^m \times \mathbb{R}^p \to \mathbb{R}$:

$$L(x,\pi,\lambda) = f(x) + \sum_i \pi_i g_i(x) + \sum_j \lambda_j h_j(x)$$

$\pi_i$ is the Lagrange multiplier or dual variable associated with *inequality* constraint $g_i(\boldsymbol{x}) \leqslant 0$, and $\lambda_j$ the Lagrange multiplier or dual variable associated with equality constraint $h_j(\boldsymbol{x}) = 0$. Note that *inequality* multipliers $\pi_i$ must be non-negative, while *equality* multipliers $\lambda_j$ can take values in all $\mathbb{R}$.

The dual function $\omega(\pi, \lambda)$ of the problem (A5.4) returns the minimum cost of a relaxed problem version, for particular values of the multipliers ($\pi \geqslant 0, \lambda$):

$$\omega(\pi, \lambda) = \min_{x \in S}\left\{ f(\boldsymbol{x}) + \sum_i \pi_i g_i(\boldsymbol{x}) + \sum_j \lambda_j h_j(\boldsymbol{x}) \right\} = \min_{x \in S} L(\boldsymbol{x}, \pi, \lambda)$$

The value $\omega(\pi, \lambda)$ is usually called the dual cost or relaxed cost associated with multipliers $(\pi, \lambda)$. The domain of the dual function (dom $\omega$) consists of those multipliers ($\pi \geqslant 0, \lambda \in \mathbb{R}^p$) such that the minimum $\min_{x \in S} L(\boldsymbol{x}, \pi, \lambda)$ exists.

### A5.4 Optimality Conditions

In this section, we introduce the necessary conditions and sufficient conditions for optimality under different assumptions. This family of conditions is usually named Karush-Kuhn and Tucker conditions, or KKT conditions for short.

#### A5.4.1 Optimality conditions in problems with strong duality

Let $(\boldsymbol{\pi}^*, \boldsymbol{\lambda}^*)$ be a maximum of the dual function of a problem (A5.4) with strong duality, and $x^*$ a primal optimal solution. Then, it holds that:

$$\begin{aligned} f(\boldsymbol{x}^*) = \omega(\boldsymbol{\pi}^*, \boldsymbol{\lambda}^*) &= \min_{x \in S}\left\{ f(\boldsymbol{x}) + \sum_i \pi_i^* g_i(\boldsymbol{x}) + \sum_j \lambda_j^* h_j(\boldsymbol{x}) \right\} \\ &\leqslant f(\boldsymbol{x}^*) + \sum_i \pi_i^* g_i(\boldsymbol{x}) + \sum_j \lambda_j^* h_j(\boldsymbol{x}) \\ &\leqslant f(\boldsymbol{x}^*) \end{aligned} \quad \text{(A5.5)}$$

The first equality in (A5.5) is met because of strong duality property. The next equality in the first line is the definition of the dual function. The first inequality holds since the minimum in a set $S$ is lower or equal than the value in a particular point $\boldsymbol{x}^* \in S$. The last inequality holds since $\pi_i^* g_i(\boldsymbol{x}^*) \leqslant 0$ (as $\pi_i \geqslant 0$ and $g_i(\boldsymbol{x}^*) \leqslant 0$) and $h_j(\boldsymbol{x}^*) = 0$. Then, the two inequalities are equalities if and only if the problem has strong duality. This leads to the following proposition.

**Proposition 5.4.1** For a problem (A5.4), we assume that there exists a primal solution $\boldsymbol{x}^*$, and the problem dual function has a maximum in $(\pi_i \geqslant 0, \lambda^*)$. Then, strong duality holds for the problem, if and only if:

- *Lagrange minimization*: $\boldsymbol{x}^*$ is a minimizer of

$$\min_{x \in S}\left\{ f(\boldsymbol{x}) + \sum_i \pi_i^* g_i(\boldsymbol{x}) + \sum_j \lambda_j^* h_j(\boldsymbol{x}) \right\}$$

- *Complementary slackness*: $\pi_i^* g_i(\boldsymbol{x}^*) = 0, \forall i = 1, 2, \ldots, m$.

Complementary slackness conditions appear only for inequality constraints. They mean that, if $\boldsymbol{x}^*$ is an optimal solution and $\pi_i^*$ an optimum multiplier for a constraint $g_i(\boldsymbol{x}^*) \leqslant 0$, then:

- If the multiplier $\pi_i^* > 0$, then the constraint is satisfied with equality in the optimum ($g_i(x^*)=0$). We say that the constraint is tight or active.
- If a constraint is inactive, or loose ($g_i(x^*)<0$), then its associated multiplier must be zero $\pi_i^* = 0$.

**Proposition 5.4.1** is the root of the derivations of the KKT conditions elaborated on in next propositions.

**Proposition 5.4.2** (KKT necessary and sufficient conditions) Let (A5.4) be a problem with strong duality, and which has at least an optimal solution, with a finite cost. Then, $x^*$ is an optimal solution of (A5.4) and ($\pi^*$, $\lambda^*$) is an optimal solution of the dual problem, if and only if all of the following conditions hold:

- Primal feasibility: $x^*$ satisfies all constraints in (B.5)
- Dual feasibility: $\pi^* \geqslant 0$
- Lagrange minimization: $x^*$ minimizes

$$\min_{x \in S} \left\{ f(x) + \sum_i \pi_i^* g_i(x) + \sum_j \lambda_j^* h_j(x) \right\}$$

- Complementary slackness: $\pi_i^* g_i(x^*) = 0$, $\forall_i = 1,2,\dots,m$.

If we apply the KKT conditions in a convex differentiable problem where all the constraints are relaxed ($S=\mathbb{R}^n$), we have:

**Proposition 5.4.3** (KKT necessary conditions for convex differentiable problems). Let (A5.4) be a problem with strong duality, and which has at least an optimal solution, with finite cost. $f$, $g_i$ are convex and differentiable, $h_j$ is linear, and $S=\mathbb{R}^n$. Then, $x^*$ is an optimal solution of (A5.4) and ($\pi^*$, $\lambda^*$) is an optimal solution of the dual problem, if (a) $x^*$ is feasible, (b) $\pi \geqslant 0$, (c) complementary slackness holds, and:

$$\nabla f(x^*) + \sum_i \pi_i^* \nabla g_i(x^*) + \sum_j \lambda_j^* \nabla h_j(x^*) = 0 \tag{A5.6}$$

The following property can be easily derived by applying KKT conditions in **Proposition** 5.4.2, in the case when no constraints are relaxed.

**Proposition 5.4.4** In the problem $\min_{x \in S} f(x)$, where $f$ is a continuously differentiable convex function and $S$ is a non-empty, closed convex set, $x \in S$ is the global minimum if and only if:

$$(y-x)^T \nabla f(x) \geqslant 0, \forall y \in S$$

The following proposition sets some conditions to guarantee that the dual optimum is unique.

**Proposition 5.4.5** (Uniqueness of the primal-dual optimal pair). Under the assumptions of **Proposition 5.4.3**, if the objective function $f$ is strictly convex, and $x^*$ is the unique global optimum. Then, if the gradients of the active constraints in the optimum are linearly independent ($x^*$ is then called a *regular point*), the optimal dual multipliers ($\pi^*$, $\lambda^*$) are unique. Note that the active constraints are all equality constraints and inequality constraints for which $g_i(x^*)=0$.

## A5.4.2 Optimality conditions in problems without strong duality

According to **Proposition 5.4.1**, if strong duality does not hold for a problem, Lagrange minimization and/or complementary slackness conditions do not hold for it either. Then, there is no primal-dual pair that satisfies KKT optimality conditions. The maximization of the dual function can help to find approximate primal solutions. This remark follows the proposition below.

**Proposition 5.4.6** For any $(\pi \geqslant 0, \lambda)$ multipliers of the problem (A5.4), not necessarily dual optimal, any minimizer $x^* \in \chi^*(\pi, \lambda)$ is a global optimum of the perturbed problem (where the right-hand side of the constraints is modified):

$$\min_{x} f(x), \quad \text{subject to:}$$
$$g_i(x) \leqslant g_i(x^*), i=1,2,\ldots,m$$
$$h_j(x) \leqslant h_j(x^*), j=1,2,\ldots,p$$
$$x \in S$$

This property can be used in different manners. We can search for the optimal dual solution, which is always a convex program. For any minimizer of the dual optimum $x^*$, it holds that $g_i(x^*)$ is the coordinate of a subgradient and for those constraints with a non-zero multiplier, this subgradient tends to be small. Then, even if the minimizers associated are unfeasible for the original problem, they will be at the global optimum for perturbed problems that can be very similar to the original one (although this is not guaranteed). This could be useful when we have some margin to accept unfeasible solutions.

# References

ABOU-SAYED A. Data mining applications in the oil and gas industry [J]. Journal of Petroleum Technology, 2012, 64 (10): 88-95.

ALBASHRAWI M. Detecting financial fraud using data mining techniques: a decade review from 2004 to 2015 [J]. Journal of Data Science, 2016, 14 (3): 553-570.

AMANI F A, FADLALLA A M. Data mining applications in accounting: a review of the literature and organizing framework [J]. International Journal of Accounting Information Systems, 2017, 24: 32-58.

ANIK M A H, SADEEK S N, HOSSAIN M, et al. A framework for involving the young generation in transportation planning using social media and crowd sourcing [J]. Transport Policy, 2020, 97: 1-18.

ATRIS A M. Assessment of oil refinery performance: application of data envelopment analysis-discriminant analysis [J]. Resources Policy, 2020, 65: 101543.

BEN-HUR A, HORN D, SIEGELMANN H T, et al. Support vector clustering [J]. Journal of Machine Learning Research, 2001, 2: 125-137.

BOGAERT M, LOOTENS J, VAN DEN POEL D, et al. Evaluating multi-label classifiers and recommender systems in the financial service sector [J]. European Journal of Operational Research, 2019, 279 (2): 620-634.

CHANDOLA V, KUMAR V. Summarization-compressing data into an informative representation [J]. Knowledge and Information Systems, 2007, 12 (3): 355-378.

CHEN H, CHIANG R H L, STOREY V C. Business intelligence and analytics: from big data to big impact [J]. MIS Quarterly, 2012: 1165-1188.

CYBENKO G. Approximation by superpositions of a sigmoidal function [J]. Mathematics of Control, Signals and Systems, 1989, 2 (4): 303-314.

DESAI J N, PANDIAN S, VIJ R K. Big data analytics in upstream oil and gas industries for sustainable exploration and development: a review [J]. Environmental Technology & Innovation, 2020: 101186.

ESTIVILL-CASTRO V. Why so many clustering algorithms: a position paper [J]. ACM SIGKDD Explorations Newsletter, 2002, 4 (1): 65-75.

FELFERNIG A, ISAK K, SZABO K, et al. The VITA financial services sales support environment [C] //Proceedings of the national conference on artificial intelligence. Menlo Park, CA; Cambridge, MA; London; AAAI Press; MIT Press; 1999, 2007, 22 (2): 1692.

GÜRBÜZ F, TURNA F. Rule extraction for tram faults via data mining for safe transportation [J]. Transportation Research Part A: Policy and Practice, 2018, 116: 568-579.

HAN J, KAMBER M, Pei J. Data mining concepts and techniques-third edition [M]. Waltham: Morgan Kaufmann, 2012.

HAN J, MORAGA C. The influence of the sigmoid function parameters on the speed of backpropagation learning [C]. International Workshop on Artificial Neural Networks, 1995, Springer.

HANGA K M, KOVALCHUK Y. Machine learning and multi-agent systems in oil and gas industry applications: a survey [J]. Computer Science Review, 2019, 34: 100191.

HARRY Z. The optimality of naive bayes [C] //FLAIRS2004 conference. 2004.

HASSANI H, HUANG X, SILVA E. Digitalisation and big data mining in banking [J]. Big Data and Cognitive Computing, 2018, 2 (3): 1-13.

HINTON G, DENG L, YU D, et al. Deep neural networks for acoustic modeling in speech recognition: The shared views of four research groups [J]. IEEE Signal Processing Magazine, 2012, 29 (6): 82-97.

HIPP J, GÜNTZER U, NAKHAEIZADEH G. Algorithms for association rule mining—a general survey and comparison [J]. ACM SIGKDD Explorations Newsletter, 2000, 2 (1): 58-64.

HONG J, TAMAKLOE R, PARK D. Application of association rules mining algorithm for hazardous materials transportation crashes on expressway [J]. Accident Analysis & Prevention, 2020, 142: 105497.

HOTHORN T, HORNIK K, ZEILEIS A. Unbiased recursive partitioning: a conditional inference framework [J]. Journal of Computational and Graphical Statistics, 2006, 15 (3): 651-674.

ISLAM M S, HASAN M M, WANG X, et al. A systematic review on healthcare analytics: application and theoretical perspective of data mining [C] //Healthcare. Multidisciplinary Digital Publishing Institute, 2018, 6 (2): 54.

KARGARI M, SEPEHRI M M. Stores clustering using a data mining approach for distributing automotive spare-parts to reduce transportation costs [J]. Expert Systems with Applications, 2012, 39 (5): 4740-4748.

KASS G V. An exploratory technique for investigating large quantities of categorical data [J]. Journal of the Royal Statistical Society: Series C (Applied Statistics), 1980, 29 (2): 119-127.

KOTSIANTIS S, KANELLOPOULOS D. Association rules mining: a recent overview [J]. GESTS International Transactions on Computer Science and Engineering, 2006, 32 (1): 71-82.

KRAUS M, FEUERRIEGEL S, OZTEKIN A. Deep learning in business analytics and operations research: models, applications and managerial implications [J]. European Journal of Operational Research, 2020, 281 (3): 628-641.

KUM H C, DUNCAN D F, STEWART C J. Supporting self-evaluation in local government via knowledge discovery and data mining [J]. Government Information Quarterly, 2009, 26 (2): 295-304.

KŮRKOVÁ V. Kolmogorov's theorem and multilayer neural networks [J]. Neural Networks, 1992, 5 (3): 501-506.

LIU Y, LI J, MING Z, et al. Domain-specific data mining for residents' transit pattern retrieval from incomplete information [J]. Journal of Network and Computer Applications, 2019, 134: 62-71.

MA X, WANG Z, ZHOU S, et al. Intelligent healthcare systems assisted by data analytics and mobile computing [C] //2018 14th International Wireless Communications & Mobile Computing Conference (IWCMC). IEEE, 2018: 1317-1322.

MAHESHWARI A, DAVENDRALINGAM N, DELAURENTIS D A. A comparative study of machine learning techniques for aviation applications [C]. 2018 Aviation Technology, Integration, and Operations Conference, 2018: 3980.

MANESSI F, ROZZA A. Learning combinations of activation functions [C]. 24th International Conference on Pattern Recognition (ICPR), 2018, IEEE.

MCGHIN T, CHOO K K R, LIU C Z, et al. Blockchain in healthcare applications: research challenges and opportunities [J]. Journal of Network and Computer Applications, 2019, 135: 62-75.

MEYER D, LEISCH F, HORNIK K. The support vector machine under test [J]. Neurocomputing, 2003, 55 (1-2): 169-186.

MEYER G, ADOMAVICIUS G, JOHNSON P E, et al. A machine learning approach to improving dynamic decision making [J]. Information Systems Research, 2014, 25 (2): 239-263.

MILJANOVIC M. Comparative analysis of recurrent and finite impulse response neural networks in time series prediction [J]. Indian Journal of Computer Science and Engineering, 2012, 3 (1): 180-191.

MISHRA V, MISHRA T K, MISHRA A. Algorithms for association rule mining: a general survey on benefits and drawbacks of algorithms [J]. International Journal of Advanced Research in Computer Science, 2013, 4 (3): 155-159.

MONTIEL L, DIMITRAKOPOULOS R. Optimizing mining complexes with multiple processing and transportation alternatives: an uncertainty-based approach [J]. European Journal of Operational Research, 2015, 247 (1): 166-178.

MOSTAFA M M, EL-MASRY A A. Citizens as consumers: profiling e-government services' users in Egypt via data mining techniques [J]. International Journal of Information Management, 2013, 33 (4): 627-641.

NGAI E W T, HU Y, WONG Y H, et al. The application of data mining techniques in financial fraud detection: a classification framework and an academic review of literature [J]. Decision Support Systems, 2011, 50 (3): 559-569.

NGAI E W T, XIU L, CHAU D C K. Application of data mining techniques in customer relationship management: a literature review and classification [J]. Expert Systems with Applications, 2009, 36 (2): 2592-2602.

PARK S, XU Y, JIANG L, et al. Spatial structures of tourism destinations: a trajectory data mining approach leveraging mobile big data [J]. Annals of Tourism Research, 2020, 84: 102973.

PERNG Y H, CHANG C L. Data mining for government construction procurement [J]. Building Research & Information, 2004, 32 (4): 329-338.

PABLO PAVóN MARIñO. Optimization of computer networks-modeling and algorithms: a hands-on approach [M]. Chichester: John Wiley & Sons, Ltd, 2016.

QUINLAN J R. Induction of decision trees [J]. Machine Learning, 1986, 1 (1): 81-106.

RALHA C G, SILVA C V S. A multi-agent data mining system for cartel detection in Brazilian government procurement [J]. Expert Systems with Applications, 2012, 39 (14): 11642-11656.

REN J, ZHANG Q, LIU F. Analysis of factors affecting traction energy consumption of electric multiple unit trains based on data mining [J]. Journal of Cleaner Production, 2020, 262: 121374.

RICO M T S. Data mining, optimization and simulation tools for the design of intelligent transportation systems [D]. Universidad de Castilla-La Mancha, 2015.

ROKACH L, MAIMON O. Top-down induction of decision trees classifiers-a survey [J]. IEEE Transactions on Systems, Man, and Cybernetics, Part C (Applications and Reviews), 2005, 35 (4): 476-487.

ROSENBLATT F. The perceptron: a probabilistic model for information storage and organization in the brain [J]. Psychological Review, 1958, 65 (6): 386.

SCHUBERT E, SANDER J, ESTER M, et al. DBSCAN revisited, revisited: why and how you should (still) use DBSCAN [J]. ACM Transactions on Database Systems (TODS), 2017, 42 (3): 1-21.

SENGUPTA S, BASAK S, SAIKIA P, et al. A review of deep learning with special emphasis on architectures, applications and recent trends [J]. Knowledge-Based Systems, 2020, 194: 105596.

SHAW M J, SUBRAMANIAM C, TAN G W, et al. Knowledge management and data mining for marketing [J]. Decision Support Systems, 2001, 31 (1): 127-137.

SHEARER C. The CRISP-DM model: the new blueprint for data mining [J]. Journal of Data Warehousing, 2000, 5 (4): 13-22.

SHRESTHA A, MAHMOOD A. Review of deep learning algorithms and architectures [J]. IEEE Access, 2019, 7: 53040-53065.

SHUMWAY R H, STOFFER D S. Time series analysis and its applications: with R examples [M]. New York: Springer, 2011.

SLOBOGIN C. Government data mining and the fourth amendment [J]. the University of Chicago Law Review, 2008, 75 (1): 317-341.

SOKAL R R, MICHENER C D, KANSAS U O. A statistical method for evaluating systematic relationships [M]. Lawrence: University of Kansas, 1958.

STROBL C, MALLEY J, TUTZ G. An introduction to recursive partitioning: rationale, application, and characteristics of classification and regression trees, bagging, and random forests [J]. Psychological Methods, 2009, 14 (4): 323.

SYSOEV A, KHABIBULLINA E, KADASEV D, et al. Heterogeneous data aggregation schemes to determine traffic flow parameters in regional intelligent transportation systems [J]. Transportation Research Procedia, 2020, 45: 507-513.

VERCELLIS C. Business intelligence: data mining and optimization for decision making [M]. New York: Wiley, 2009.

WITTEN I H, FRANK E, HALL M A, et al. Practical machine learning tools and techniques [J]. Morgan Kaufmann, Burlington, MA, 2011, 10: 1972514.

XU D, TIAN Y. A Comprehensive survey of clustering algorithms [J]. Annals of Data Science, 2015, 2 (2): 165-193.